KOREA

KOREA

A CARTOGRAPHIC HISTORY

JOHN RENNIE SHORT

THE UNIVERSITY OF CHICAGO PRESS
CHICAGO AND LONDON

KOREA **KF** FOUNDATION

The Korea Foundation has provided financial assistance
for the undertaking of this publication project.

John Rennie Short is professor of public policy at the
University of Maryland (UMBC).

The University of Chicago Press, Chicago 60637
The University of Chicago Press, Ltd., London
© 2012 by The University of Chicago
All rights reserved. Published 2012.
Printed in China

21 20 19 18 17 16 15 14 13 12   1 2 3 4 5

ISBN-13: 978-0-226-75364-5 (cloth)
ISBN-10: 0-226-75364-6 (cloth)

Library of Congress Cataloging-in-Publication Data

Short, John R.
    Korea : a cartographic history / John Rennie Short.
        p. cm.
    Includes bibliographical references and index.
    ISBN-13: 978-0-226-75364-5 (cloth : alk. paper)
    ISBN-10: 0-226-75364-6 (cloth : alk. paper)
    1. Cartography—Korea—History. I. Title.
    GA1251.S56 2012
    912.519—dc23

                                        2011021062

♾ This paper meets the requirements of ANSI/NISO
Z39.48-1992 (Permanence of Paper).

For my father, John Ingles Short (1933–1997)

CONTENTS

Preface   ix

1 Introduction: The Globalization of Space   1

PART 1   SEPARATE WORLDS

2 Early Joseon Maps   13
3 Europe Looks East   32

PART 2   CARTOGRAPHIC ENCOUNTERS

4 Joseon and Its Neighbors   51
5 Cartographies of the Late Joseon   77

PART 3   REPRESENTING KOREA IN THE MODERN ERA

6 The Colonial Grid   113
7 Representing the New Country   127
8 Cartroversies   137

Guide to Further Reading   153
References   157

The origins of this book lie in my abiding interest in the history of cartography. Maps have always fascinated me. My love of maps impelled me to study geography at university. Yet by the time I went to the University of Aberdeen in 1969, the discipline had moved past its early concern with maps. The new emphasis was on spatial modeling and broader social issues. These topics also interested me, so my love affair with old maps remained dormant as I followed the dominant themes of quantitative rigor and social criticism. It was only later in my career that I revisited maps as objects of study and analysis. Concurrently my academic discipline was rediscovering and renewing its interest in the history of maps. Brian Harley in particular was promoting a more critical engagement (Harley 2002). Influenced by this postmodern school of thought, I wrote a number of books that presented the cartographic representation of the United States (*Representing the Republic*, 2001), the development of a spatial sensitivity in the early modern period in Europe (*Making Space*, 2004), and the role of indigenous peoples in mapmaking (*Cartographic Encounters*, 2009), as well as a general book on the history of cartography (*The World through Maps*, 2003). This current book is a continuation of my lifelong and incurable cartophilia.

This book also arose out of my longtime interest in Korea. I first learned about the country at an early age from the stories my father told about his experiences as a young man sailing halfway around the world to fight in the Korean War. It was a life-changing experience for him. The war's harrowing events in this far-off land stayed with him for the rest of his life as he sought, unsuccessfully, to make sense of them. My interest in the country is an enduring link with my father, who faced enormous danger in Korea while on the other side of the world, in Scotland, his eldest son was born. When I visited South Korea for the first time in 1993, I went to some of the same places he had visited many years before in such different circumstances. He was there at a time of intense conflict, in a world very much divided. I was there at a time of relative peace. I made friends in South Korea, had some of my books published in the Korean language, and began a long association with Korean geographers that continues to the present.

Those two broad interests—maps and Korea—were brought together at the prompting of my old friend and colleague Ki-suk Lee. It was he who first suggested I give a presentation at the Fourteenth International Seminar on Sea Names held in Tunis, Tunisia, in August 2008. I delivered a paper on the geopolitics of naming seas that provoked my interest in the cartographic representation of Korea. My proposal to write a book was enthusiastically accepted by the Korea Foundation, which graciously provided financial assistance to allow me to travel to libraries around the world and take the time necessary to complete the manuscript.

There is an existing literature on cartographic representation of Korea. Standard cartobibliographies include Lee's *Old Maps of Korea* ([1977] 1991). In English, the most comprehensive work to date on the history of the cartography of Korea is Ledyard (1994). Han, Ahn, and Sung (2008) provide a concise, beautifully illustrated description of Joseon cartography, drawing primarily on the Kyujanggak Collection–the royal library of the Joseon dynasty-at Seoul National University. There are also more detailed studies such as the articles that appeared in the Spring 2008 issue *of Korea Journal*. There are, however, no studies that deal either with the totality of Korean cartographic representation or specifically with the Western cartographic representation of Korea, and none that are easily accessible to the non-Korean audience.

Another broad interest of mine is globalization. I have written a number of books detailing its specific effects on cities (*Globalization and the City*, 1999; *Global Metropolitan*, 2004) as well as outlining its spatial impacts and consequences (*Global Dimensions: Space, Place, and the Contemporary World*, 2001). I am fascinated by the production of global space, which I take to have two distinct meanings: creating an understanding of the world's geography and representing this geography in cartographic form. A traditional view sees both things simply as a Eurocentric process in which European powers incorporated the world into a global discourse and created new, more modern forms of cartographic representation. It is the story of how Europe explored, discovered, and mapped the world, displacing older indigenous cartographies with more scientific mappings. Modernity in this rendering is exported from Europe to the rest of the world. There is now an exciting body of work that points to the creation of new knowledge in the process of interactions and encounters between the colonial and the colonized, the imperial center and the periphery, the modern and the indigenous. Knowledge was not so much imposed from one region of the world as created and disseminated in a complex series of interactions in many regions. Modernity was created in the space of encounters, continually imported, transformed, and modified, then reexported in a continual and ongoing global circulation of ideas and practices. Here I will explore this general notion of the production of global space and the more specific theme of complex global cartographic encounters in understanding the history of Korean cartography.

This book is a general introduction to how Korea was and is represented in maps. Part 1, "Separate Worlds," reveals the differing cartographic traditions prevalent in the early Joseon period in Korea and its temporal equivalent in early modern Europe, roughly from 1400 to 1600. In Joseon Korea a sophisticated cartography drew on Chinese influences, themselves drawing on Arabic and European knowledge to envision the world. In chapter 2 I show how the emphasis of the early Joseon was on mapping the nation-state and its near neighbors. Maps were an important form of surveillance and a vital information base for ensuring political control and maintaining regime legitimacy. A rich variety of pictorial styles developed as the state mapped its territory and surveyed its borders. At the same time, European merchants and explorers were traveling to the region. In chapter 3 I look at the cartographic emergence of Korea in early European maps. This production of global knowledge was not so much a cartographic imposition as a cartographic encounter between Europe and East Asia. An improved understanding of Korea for Europeans and the discovery of Europe for Koreans became part of a new global perspective. Edward Said (1978) described the practice of defining the East by those in the West as *orientalism*. The process is always more complex than Said first theorized, and I want to explore this complexity by looking at how Korean mapmakers embodied, reflected, and even contested Western depictions. A number of similar projects have been conducted for both China and Japan. These include Min-min Chang's *China in European Maps* (2003) and Lutz Walter's *Japan: A Cartographic Vision* (1994). There is no Korean equivalent, and that major gap in our understanding does a disservice to the global appreciation of the Korean role in world intellectual history.

The discussion of cartographic encounters between Korea and the rest of the world is central to part 2, "Cartographic Encounters," which covers the period roughly from 1600 to 1900. Chapter 4 explores how a distinctly Korean cartography is in fact a product of encounters with impinging empires and other nations and centers of representation. The earliest influence was China, but I show the influence of Japan and Europe as well. I also show how the West became known to the Koreans and the cartographic effects resulting from this intellectual encounter. Joseon rule encompassed an enormous range and depth of cartographic production. I make a distinction between the early and late Joseon, and in chapter 5 I highlight the cartographic connections of the late Joseon with its neighbors, spending some time providing a context for the masterpiece of late Joseon cartography, the 1861 *Daedong yeojido* (Map of the Great East [Korea]). This work is a culmination of the encounter between an indigenous Korean cartography and a modern, more universal cartographic practice. This hybrid map, neither simply Korean nor decontextualized modern but a subtle combination and intermingling of the two, stands as an example of two centuries of cartographic encounters. It is a masterpiece that uses modern methods and indigenous practices to create the best-known "modern Korean" map of Korea, combining "Korean" and "modern" in one artifact. It is probably the best example of "early modern Korean." Chapter 5 details how the ending of the dynasty's long seclusion in the late nineteenth century is recorded in map inscriptions.

Part 3, "Representing Korea in the Modern Era," covers the period from Japanese colonization of the country to the present. I will demonstrate how some of the tumultuous events of the past 120 years are recorded and contested in maps. Chapter 6 shows how Japanese mapmakers incorporated Korea. This cartographic encounter shows marked asymmetry, a form of Japanese colonial control. Chapter 7 considers postcolonial Korea and the cartographic implications of the continuing North/South split. Chapter 8 covers recent "cartroversies" of the national representation of the peninsula, the naming of the East Sea/Sea of Japan, and claims of ownership of the island of Dokdo. Maps, ancient and modern, play an important part in these contemporary debates.

Writing a book in English about a Korean topic is fraught with linguistic pitfalls. The romanization of Korean has gone though a number of recent changes. The revised romanization adopted by the government of South Korea in 2000 was introduced to be more amenable for computer use, dispensing with apostrophes, breves, and diacritics. This system replaced the modified McCune-Reischauer system that was in use from 1984 to 2000, which in turn modified a system by the same name first introduced in 1937. The three systems can be found in even the most recent publications. The dynasty that ruled Korea for over five hundred years is thus written in English as *Joseon* under the revised system, *Chosun* under the modified McCune-Reischauer, and *Yi* in the original McCune-Reischauer. When we get to mapmakers it is important to remember that Kim Jeong-ho, Kim Chong-ho, and Gim Jeong-ho are the same person. Not only are different systems used, but sometimes they are used creatively. Many texts, for example, use the McCune-Reischauer system but omit the breves and apostrophes. I have tried to adhere to the more recent revised romanization form, but not always. For example, Kim Jeong-ho is not the most recent form, which is Gim Jeong-ho, but it is the most widely used and best-known variant. Although "a foolish consistency," as Emerson reminds us, "is the hobgoblin of little minds," I still ask readers' indulgence for the inconsistencies. In time periods, all undesignated dates are in the Common Era. Dates that are Before the Common Era are specified as BCE. In later chapters I use the more common terms North and South Korea to refer respectively to the Democratic People's Republic of Korea and the Republic of Korea.

I have a very specific audience in mind for this book. It is not meant as a text for Korea experts or for historians of cartography. While I hope I have something interesting to say to both groups, the book is written very much for a general audience, with little or no facility in the Korean language, perhaps coming to an understanding of the country and of maps for the first time.

Maps are a subtle form of communication that tell us much about their makers, about when they were made, and about that time. Maps are an invaluable record of how certain groups envisioned, represented, and saw places and locations. Maps by necessity are selective partial views, but the selection of what is important and what to leave out itself tells us much about the worldview of the maker as well as the readers of the map. Maps do not reflect a general reality so much as they embody particular realities made by particular people at specific times in specific places. But maps are slippery witnesses that often carry conflicting and deeper meanings. In this book I will question these witnesses with a view to uncovering their deeper truths and evasive answers.

I have cited only the scholarly literature in English. Korean language texts, while of great value, are inaccessible to those unable to read Korean. My ideal reader is someone like my father, who tried to make sense of a country very different from his own. This introductory history of Korean cartographic representation is, as I realized only after I completed the text, a gift to him for all the things unsaid and all the things not done. I miss him more as the years pass.

1

INTRODUCTION
THE GLOBALIZATION OF SPACE

Nations are embodied in many things, including ruling dynasties, capital cities, cuisine, language, dress, sports, customs, and stories. In this book I will look at just one object of material culture that expresses national identity—the map. I want to consider how Korea was represented in and through maps over the six hundred years from the end of the fourteenth century to the present day.

Korea, once known as the Hermit Kingdom, is now an important part of the modern world. Throughout the twentieth century its story was significant in world history. In the early twentieth century its annexation by Japan was one example of the seizing of smaller countries by more powerful entities, and what befell Korea then is an integral part of the global colonial experience. In the post–World War II era, its division into North and South reflected the deeper geopolitical division between East and West. The Korean War was a major flashpoint of this divided world. From hostilities breaking out in 1950 to the Armistice in 1953, military and civilian casualties and deaths are estimated at between 4.2 and 4.7 million. The Armistice established the border roughly at the thirty-eighth parallel, with a Demilitarized Zone stretching two kilometers on each side. A formal peace treaty between North and South Korea has still to be signed. The cease-fire froze relations on a permanent war footing, and the national division between North and South endures and has widened while the East/West global rupture has long since healed. Although many communist regimes collapsed in 1989, North Korea remains a nuclear threat. Meanwhile, South Korea has emerged as an Asian tiger of rapid and sustained economic growth fueled by a dynamic export-led manufacturing sector. South Korea is now ranked the eighth largest exporter and fifteenth largest economy in the world as measured by gross domestic product (GDP). While South Korea joins the ranks of high-income developed economies, North Korea teeters on the edge of economic collapse, with most of its population facing recurring famine and economic deprivation. Their divergent fortunes are starkly revealed by the raw economic data: the South has a GDP of $30,200 per capita; North Korea's is $1,900.

The Korean peninsula remains an important part of the world at the beginning of the twenty-first century. South Korea is a major economic power, and North Korea continues to exercise political leverage and provoke fear among its near neighbors and distant superpowers. To understand Korea is to understand something vital to the history and geography of the modern world.

I will use maps as the main vehicle to promote such an understanding. I will show how Korea was represented in maps both from within and from without and discuss how these processes and the resulting interaction were always contested. Maps are a unique historical intersection of culture, technology, and political considerations. They provide a narrow aperture that allows a deep entry into the history and geography of the country.

A map is not a simple thing. It combines technical considerations with social and political messages. Maps have multiple meanings: they have symbolic importance; they are decorative; they have practical applications as well as ideological underpinnings. Maps are complicated texts used for a variety of purposes and read by diverse readers. They embody technical progress, social development, and political conflict. More than mere depictions of territory, they are political statements and discursively subtle social arguments that tell us much about society.

To use a term such as "Korean cartography" is perhaps to overstate the singularity of a unique cartographic tradition. Korean maps drew from non–Korean sources, particularly Chinese, but also—despite Korea's reputation as a relatively closed society until the late nineteenth century—from a range of other cartographic influences. Korean maps are more accurately referred to as "Korean but with many and diverse foreign influences." Throughout this book, the terms "Korean maps" and "Korean cartography" should replace yet evoke this awkward phrasing. Korean maps are a result of diverse cartographic encounters.

Maps of nations arise from internal and external forces. The need for central rulers to plot their kingdom intermingles with outsiders' need to identify and locate another land. There is a complex evolution in the mapping of national spaces, shaped by local and global mapmaking requirements. The maps of each country result from the complex intermingling of local, national, and global cartographic perceptions that are shaped by interaction with each other. The long gradual shift in Korean cartography, from a fourteenth-century Sinocentric perspective to a more international twentieth-century one, is part of the broader and deeper history of the country's involvement in global discourses, in the production of global space, and in the making of the modern world.

The evolution of Korean cartography is part of the production of global space, the processes through which an understanding of the world's geography emerges and a universal mapping project is enacted. Charles Parker (2010) writes of the global integration of space as the defining feature of early modernity. From 1400 to 1800, empire building across the globe—including the Chinese, Ottoman, Mughal, and Safavid empires as well as those of Spain and Britain—created international markets and global exchange networks, the movement of people, the spread of new technologies, the diffusion of cultures, and the transmission of scientific practices and religion. The result was a tighter integration of global space. Parker shows how increasing contact between cultures led to cross-cultural borrowings. In the case of cartography he highlights, among others, the encounters between Chinese, Ottoman, and European mapmakers. The Ottoman cartographer Piri Reis (1470–1554) made a world map in 1513 that drew on Arab, Chinese, and Iberian cartography. In a detailed case study of the cross-cultural and transnational creation of science and scientific practices, Kapil Raj (2007) describes the integration in the circulation of knowledge between South Asia and Europe from 1650 to 1900. Raj undermines the traditional view that science was exported from the European core to the colonial periphery, showing that an intercultural encounter produced new knowledge. In particular, the geographical exploration of British India in the late eighteenth and early nineteenth centuries provides a good illustration of the way British and Indian practitioners and skills met around specific projects, how they were reshaped, and how the modern map and its uses emerged together in India and Britain through the process of colonial encounter (Raj 2007, 63).

Local people and indigenous techniques were employed in the large-scale mapping of the continent as the British employed and redeployed local knowledge. And in the process both British and local knowledges were transformed. Encounters produced new knowledge. Laura Hostetler (2001, 2009) provides another detailed example of a cartographic encounter between the Qing dynasty in China and Jesuit mapmakers. More on this later.

The central proposition of this book is that the cartographic representation of Korea is part of the general story of production of global space, and in this particular case the result of a series of wider cartographic encounters between Asia and Europe in which the foreign and the indigenous mapmaking traditions are continually interacting to produce new hybrid forms. The evolution of Korean cartography is part of the emergence of modern mapmaking, which itself is a result of the global integration of space, the global circulation of knowledge, imperial mappings, and a range of colonial and neocolonial encounters.

Before the integration of global space, it is easier to discern maps as embodying radically different views of the world. Let us begin, then, with two world maps produced at approximately the same time in different parts of the world, at the dawn of the early modern period. The first is one of the oldest Korean maps in existence (fig. 1.1). Commonly referred to as *Gangnido*, its full title is *Honil kangni yoktae kukto chi to* (Map of Integrated Lands and Regions of Historical Countries and Capitals). The map is drawn on silk, rolled up on a baton and meant to be unfurled for viewing. This unfurling of the world is a persistent theme in Korean cartography. The world is encoded in folded texts whose unfolding is both a reading and a display of the world. The unfurling also has the symbolic importance of revealing a secret text. For much of the long Joseon period (1392–1910) maps were government property, centralized and restricted. Civilians were not allowed to own them. Maps, especially in the early Joseon, were secret documents with access restricted to the few. The complex binding, folding, and furling were symbols of the map's power, potency, and mystery.

The *Gangnido* is a large map, 164 centimeters by 171 centimeters. Its exact date of production is unknown, but Robinson (2007) estimates the date between 1479 and 1485. This map and four others, all now in Japan, derive from an original world map made earlier, the 1402 *Gangnido*. The later *Gangnido* shown in figure 1.1 is not a simple copy of this earlier map but a revision that contains more up-to-date information on Korea. It is the oldest extant cartographic representation of the country. The map is centered on China, and Korea's size is exaggerated. China dominates the center stage, the Great Wall is shown, and the river systems are depicted in intricate patterns, more decorative than realistic. Japan is shown, albeit oriented with west at the top and positioned too far south for geographic accuracy. The rest of the world forms the periphery. In the upper left corner the rough outlines of Africa and the Arabian peninsula can be made out. India is subsumed under the Chinese continent. The map reflects the dominant Korean view of the world, where China is the center of the world and Japan is a presence, though represented slightly inaccurately as farther away from Korea than it actually is, while the rest of the map shows the outer edges of the known world.

The map also embodies the worldview of the Korean ruling elite, who needed to have accurate information about the national territory and especially to know about their immediate neighbors: China, a dominant cultural and military force in the region, and Japan, a rival state. The rest of the world shades into marginal insignificance. The map expands Korea's size and meticulously depicts its administra-

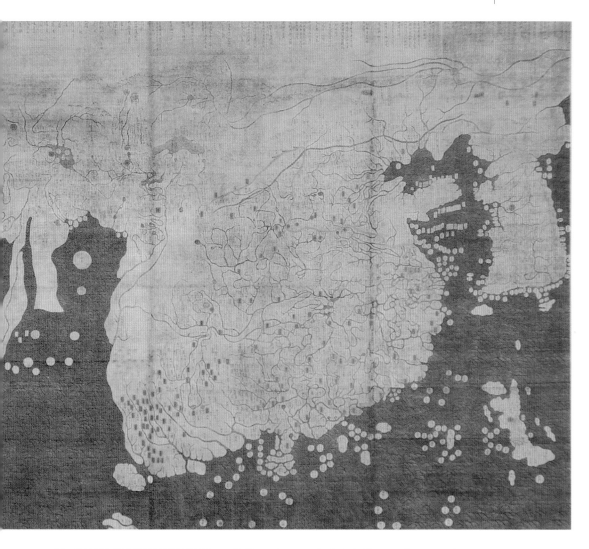

1.1 THE WORLD FROM THE EAST, c. 1480. (*Gangnido*, c. 1479–85. Ryukoku University, Kyoto, Japan.)

tive centers and military bases. Korea is drawn from more up-to-date information, while the rest of the world is depicted as it was in the 1402 base map. A contemporary Korea is embedded in an anachronistic world. A red circle surrounded by walls indicates the capital Hansong (Seoul). Similar symbols are used for the capitals of China and Japan. The main administrative centers are recorded, and naval bases are situated along the coast, since at that time Japanese sea pirates were a major threat. It is a world map with three levels of accuracy: at the center is an up-to-date picture of the administrative and military geography of Korea; in the next ring are clum-

1.2 THE WORLD FROM THE WEST, 1482. (Claudius Ptolemy, *Geographia* [Ulm: Leinhart Holle, 1482]. Courtesy of the Library of Congress, Washington, DC.)

sier, more anachronistic depictions of China and Japan; and finally the rest of the world forms a hazy outer ring.

It would be wrong, however, to see this map as only an inflated chauvinistic view of Korea's position in the world. It is interesting that Korea is placed in a world context at all. The map implies that there is a wider world—the farther from Korea, the less well known—but wider nevertheless. The second point is that the map embodies other influences. The text is in Chinese, but the map draws on the work of Islamic geographers and mapmakers whose maps were brought to China during Mongol rule. The place-names in Africa and Asia derive from Persian and Arabic originals. Even at the beginning of the early modern period, then, there is evidence of cross-cultural encounters.

The second depiction of the world is a European map made at about the same time, in Ulm in Bavaria (fig. 1.2). It is the widely recognized European worldview before Columbus set sail. This late fifteenth-century map derives from the work of Claudius Ptolemy, a Greek Egyptian who worked in the great library of Alexandria in the second century after the birth of Christ. The library was a major center of scholarship and the intellectual hub of the Hellenistic world. One of Ptolemy's major works was the *Geography*, written between 127 and 155, which developed the idea of latitude and longitude, with lines based on the older Babylonian number system of sixty. The book contains tables of latitude and longitude of sites in various parts of the known world, including Europe, Africa, and Asia. Ptolemy also made maps of Europe, four maps of Africa, and twelve of Asia. These maps did not survive, but he provided the code he used to make them. The Arab cosmologists of the Abbasid court in Baghdad translated Ptolemy's work about 800. Kept alive by Arab and Persian scholars, his reputation and accomplishments spread during the European Renaissance as his work was published and printed, copied and amended, written and read by artists, humanists, scholars, explorers, princes, priests, merchants, and prelates (Short 2004). The translation and publication of Ptolemy's *Geography* were central to the European Renaissance. The first printed copy appeared in Vicenza in 1475. It contained no maps but had two diagrams of map projections. At about the same time, other editions appeared in Rome (1477), Florence (1480), and Ulm (1482). The Ulm edition, with its painted roundels and richly colored woodblock maps including deep blue seas and yellow borders, is arguably one of the most beautiful.

The Ulm map shows the world as known to the ancient Greeks and to Europeans on the eve of the great exploration of the New World, comprising Europe, Asia, and Africa. It has yet to incorporate Portuguese exploration around the southern tip of Africa, so Africa is shown as giant landmass with the Indian Ocean as a closed sea. Since the map was made before Columbus sailed, there is nothing between the edges of western Europe and Asia. As one gets farther from the Mediterranean, the geographical knowledge declines sharply as the more accurately represented well known fades into the haziness of the unknown.

The 1482 Ulm edition world map was published almost at the same time as the *Gangnido,* yet the two embody differences, some obvious and others not so obvious. The Ulm map represented European knowledge of the world before the New World was discovered. It was reproduced in multiple copies intended for wider circulation. It was part of the broader diffusion of geographical knowledge and learning brought about by the printing and distribution of texts. The *Gangnido*, in contrast, is a manuscript, unique, with limited circulation and exposure, meant only for senior administrators. The Ulm map is part of a wider diffusion and democratization of knowledge, whereas the *Gangnido* represents elite knowledge and centralized political control. The Ptolemy map has a grid of latitude and longitude

1.3 A MODERN WORLD, 1834. (Choe Hangi, *Chigu chonhu* [Map of the Earth: Front and Behind], 1834. Museum of Sungshin Women's University, Seoul.)

that enmeshes locations in both absolute and relative position. The *Gangnido*, in contrast, has no grid. The biggest difference, of course, is that while the two maps share Persian and Arab sources, one is definitely a Western European map while the other is most definitely East Asian. Ptolemy's map runs out of accurate knowledge on its eastern edge, while the same thing happens at the western edge of the *Gangnido*. At the end of the fifteenth century the two maps reflected geographical knowledge centered in different parts of the world. Both world maps were only partial representations, clipped visions of a wider world. But both, while having a restricted vision of the world, also contained possibilities. The *Gangnido* does show the outline of Africa and Europe, and the Ulm edition does show new territories in

Scandinavia bursting through the confining frame. Both maps, partial and biased, also prefigure other worlds. In the subsequent chapters I will show how the two worldviews merged.

For the moment let us consider figure 1.3, one part of a double hemisphere woodcut map of the world that blends two previously separate traditions. The map was made by Choi Hangi (1803–75), an intellectual in the late Joseon. It draws on earlier Chinese maps that in turn were influenced by the European Jesuits working in China in the late sixteenth and early seventeenth centuries. The Jesuits introduced the double hemisphere map and a less Sinocentric view of the world. However, as Richard Smith argues, "From the late seventeenth century into the early nineteenth century, the vast majority of Chinese mapmakers ignored Jesuit constructions of the world" (Smith 1996, 59). There were exceptions. The Chinese scholar Zhuang Tingfu, for example, wrote books and produced a map in 1800. His work was brought to Korea and influenced scholars such as Choi Hangi, who were keen to adopt a more modern perspective. Zhuang Tingfu's work was the basis for Choi Hangi's 1834 map shown in figure 1.3. Notice that the world is now encased in lines of latitude and longitude, with the equator depicted as an oblique line tracking the sun's route across the seasons. And in this world Korea was represented. The map was reproduced in Choi Hangi's popular book *Chigu chonyo* (Descriptions of the Nations of the World). This figure highlights the hybridity of the late Joseon cartography, which draws on Jesuit and Chinese influences to make a "modern Korean" map of the world. Whereas figures 1.1 and 1.2 are examples of separate worldviews, a Korean map and a European map, respectively, figure 1.3 is at once both Korean and modern, the result of the complex cartographic encounter between the two worldviews that shows modernity in shared practices and in inhabiting a shared global space. Part 1 considers two worldviews that have some contact with each other, though limited. Part 2 explores their interactions.

PART

1

SEPARATE WORLDS

2 EARLY JOSEON MAPS

Although many references to maps appear in Korea before the fourteenth century, the oldest extant maps emerge at the beginning of one of the longest dynasties in world history. From 1392 to 1910 a single dynastic order ruled Korea. Based on Neo-Confucian principles imported from China, the Joseon dynasty constructed the fundamental identity of contemporary Korea: the capital was established in Seoul; the border was extended to the northernmost limit of what is now North Korea; and the distinctive Korean system of writing still in use was invented.

ORIGINS

Korea is an ancient civilization. A proto-Korean civilization, the Gojoseon, probably lasted from about 2300 to 100 BCE. Almost two thousand years ago, the three kingdoms of Goguryeo, Baekje, and Silla dominated the peninsula. One of the states, Silla, rose to preeminence, expanded dramatically between 500 and 565, exercised control over much of contemporary Korea beginning about 668, and developed strong cultural links to China. Governance styles and religions were imported from China, and Buddhism and Confucianism found their way into Korea: the Buddhist scholar and writer Wonhyo (617–86) spread the teachings of Buddhism throughout the country, and So Ch'ong (c. 660–730) taught the tenets of Confucianism. Queen

Sondok (r. 632–47) fostered close ties with China and actively promoted Buddhism. Under Queen Chindok (r. 647–54) Chinese court dress was adopted, the Chinese calendar was used, and the teaching of Chinese history and writing was encouraged. Korea was at the end of the Silk Road and thus absorbed cultural influences from India and Central Asia as well as China.

The last dynasty before Joseon was the Goryeo dynasty, ruled by thirty-four monarchs from 918 to 1392; its name, also written as Koryo, is the basis for Korea's current name. The Goryeo rulers built a new capital at Kaesong (not far from Seoul) in an imitation of the Chinese Tang architectural style. A civil service was established on the Chinese model but was less meritocratic. Buddhism flourished in rich monasteries. The printing of Buddhist texts was important. Woodblock printing was established as early as the eighth century, and the first movable metal type was used in 1377. However, this sophisticated social order was threatened as the Mongol people rose into a cohesive expansionist force. Mongols attacked Korea in 1231, and made four more major incursions from 1253 to 1258. In 1258 Korea was annexed by the Mongol empire, although Korean monarchs were allowed to rule under Mongol suzerainty. When the Korean monarch in the late fourteenth century proposed invading China, the commanding general, Yi Seong-gye, rebelled. He killed members of the elite, established himself as leader, and inaugurated the six-hundred-year Joseon dynasty that lasted from the very late fourteenth century to the very early twentieth.

The first two hundred years under Joseon rule were a time of stability and incredible cultural energy and innovation as Korea emerged from over a century of Mongol domination to experience a period described as one of "brilliant artistic creativity" (Haboush 2009). Paralleling the cultural change was the entrenchment of Neo-Confucianism as the central element of political theory and practical governance. The Joseon dynasty was one of the first societies founded on the secular ideology of Confucianism. The new order centralized power. After 1400 the state abolished private armies and limited the private ownership of weapons, thus gaining a monopoly over the means of violence. The country was divided into administrative units—provinces, counties, districts—and all provincial posts were filled with officials sent by the central government. Cultural emphasis was on civil order and political stability rather than on religious devotion.

Under early Joseon rule Korea took on many characteristics that it retains to this day. The Korean written text was invented by King Sejong (r. 1418–50), the fourth monarch of the Joseon dynasty, who codified the new written language of Hangeul about 1444. Spoken Korean is a separate language from Chinese, but because of the cultural dominance of China over Korea, the written language was based on Chinese characters. Literacy was limited to upper-class males. The new alphabet was more accessible than Chinese script; a new national language was an important part of the creation of a unified society. Hangeul let women and commoners attain literacy. According to Sejong himself, it was developed to allow the common people to express their thoughts and feelings. It was thus instrumental to a sense of Korean consciousness and identity. However, the Chinese script, hanja, remained dominant in scholarly circles and was used in most Korean maps until the nineteenth century. Hangeul was not used in official documents until 1894.

THE CONFUCIAN CONTEXT

To understand the Joseon dynasty, it is important to situate it against the wider background of Confucianism. The doctrine is named after the teacher Confucius (551–479 BCE), who lived in China. He was born during the Zhou dynasty that ruled from 1050 to 256 BCE. The rulers claimed a mandate from heaven; they were the link between heaven and earth. The Chinese character for king is one vertical line bisecting three horizontal lines; the bottom line signifies earth, the top one is heaven, and the middle one is the emperor. This character in turn forms a portion of the character for emperor. In the latter half of the dynasty various scholars and associated schools of thought sought to answer the basic question of the best form of government. Laozi (Lao Tzu) mapped out the beginnings of Daoism, which stressed minimalist intervention, or acceptance of the way of the world. Mozi (Mo Tzu) established a school of thought focusing on the notion of universal love. There was also a legalist school that preached a policy of severe punishment to shape human behavior. Confucius adopted a pragmatic position between the acceptance of Laozi, the idealism of Mozi, and the severity of the legalists. He sketched out a path for an ordered, stable polity based on traditional values of virtue, humanity, and moderation. The government's mandate was part of a wider set of idealized relationships between family members, including parent and child, husband and wife, and brothers and sisters as well as ruler and subject. In each of the relationships a set of duties entailed loyalty, devotion, and respect. Confucius never held any important political post, but his ideas would come to dominate political and moral thought throughout East Asia. His praise of scholars over soldiers and of peace over war promoted a commitment to learning and the development of the scholar-administrator political class throughout much of the region in subsequent centuries. His texts were part of the basic training of government officials up to recent years. The wider legacy is the promotion of a strong work ethic, an emphasis on the society and the family over the individual, and belief in the transformative power of learning and the arts. He created a social philosophy based on learning, devotion to public service, respect for tradition, and moderation in social behavior.

This model for government was taken up by the Han dynasty, which ruled from 220 to 206 BCE, but its collapse brought Confucianism into disfavor. In this chaos of dynastic collapse, Indian Buddhism and indigenous Daoism became more important elements of religious belief, social thought, and political belief. Buddhism flourished during the Tang dynasty (618–907) and was widely adopted in Korea. However, Confucianism was resuscitated under the Song dynasty (978–1279) in a body of thought commonly referred to as Neo-Confucianism, which is primarily concerned with cultivation of the mind, conduct of government, and the ethics of everyday life rather than with musings on abstract considerations or spiritual matters. It is a more secular model of political arrangements and social living. While one school, the Lu-Wang, stressed self-cultivation and intuition of the proper way, another approach, the Zhu Xi school, was critical of this Buddhist-influenced discourse and argued for the importance of study and investigation. This latter school was the dominant form of Neo-Confucianism in Korea.

LEGITIMIZING THE REGIME

The new Joseon order needed to be justified and legitimized. A new capital was established in Seoul, and the lavish palaces of Gyeonbokgung and Chandeokgung were built at its center in 1395 and in 1405. It is in this context of territorial stock-taking, political centralization, and establishing legitimacy that that we now see the *Gangnido* (fig. 1.1), deriving from an original world map made in 1402. The 1402 *Gangnido* map was commissioned in the very early years of the Joseon regime by two government ministers, Kim Sahyong and Yi Mu. They had visited China and had worked on maps of the northern border. The 1402 map, drawn up by Yi Hoe, an official scribe, is a composite based on many sources. The basic Sinocentric world perspective was taken from Chinese sources, including world maps made by Li Zemin and Quan Jin, and perhaps the *Da Ming hunyi tu* (Integrated Map of the Great Ming), made about 1389. The Korean section builds on a map of the country produced by Yi Hoe in early 1402. Japan's depiction derives from a Japanese map made by Gyoki and obtained by Pak Tonji, a Korean official who visited Japan between 1398 and 1402. The 1402 map places an enlarged Korea, almost as big as China, at the center of the world. It was a large map meant for public display, and it transmitted the message of a large Korea positioned next to China, the world's largest state and a cultural powerhouse. The 1402 map both placed the country next to the center of civilization and legitimized the new regime's power.

MAKING PICTURES OF THE LAND

The new Joseon dynasty not only created Korean society but also visualized it. This took many forms, including the keeping of official records and developments in landscape painting as well as formal mapmaking. The Joseon dynasty carefully recorded its rule. We still have the *Annals of the Joseon Dynasty*, a detailed record of court events in over one thousand volumes spanning 472 years. Texts, paintings, and maps document the day-to-day life of the court as well as special banquets and ceremonies. The *Annals* are full of pictures of life at court, the grand events and the great rulers as well as the more mundane events and lowly officials. The *Annals'* depiction of court life reflects the importance of visualization to Joseon rule. Things were recorded both in time and in space, and techniques of visual representation were elaborated and improved. At the Dohwaseo, the Royal Painting Institute, painting and drawing techniques were refined and improved by a cadre of skilled artists. Under early Joseon rule Korean landscape painting also developed away from Chinese influence, with its emphasis on idealized landscapes, toward new artistic forms that favored naturalism. The paintings celebrated the Korean landscape and helped foster a national consciousness. Anonymous painters as well as well-known artists such Jeong Seon (1675–1759) and Gang Se-hwang (1713–91) added to a long tradition of landscape painting that by the middle to late Joseon had developed a rich variety and deep complexity. At least eight types of brushstrokes were commonly used, including the lotus leaf, short linear, ax cut, and oxtail.

Landscape depictions were also part of the documentary paintings, *gyehadeo*, used to record and celebrate political events such as the inaugurations of provincial governors or civil service examinations. A painting from about 1621 by the painter Han Sigak, now in the National Museum of Korea, is titled *Special National Examination for Applicants from Northeastern Provinces.* Celebrating national examinations for candidates seeking to become government officials, it employs the realistic landscape style developed in the early Joseon period and visually represents the Neo-Confucian character of the state.

There are two central elements in early Korean cartography: the making of world maps that place the nation in a Sinocentric world, and the construction of national maps for the administration of the country. "By looking at maps," notes one commentary on a 1402 world map, "one can know terrestrial distances and get help in the work of government" (quoted in Ledyard 1994, 245).

The Joseon regime made many maps at all scales, from national renderings to more detailed maps of local areas, counties and towns, fortifications and harbors. Mapping was an important form of surveillance and control. At the regional and local scales, mapmaking and landscape painting went hand in hand. Artists who made paintings also made maps. Reflecting this connection, the Dohwaseo was also known as the Office of Charts and Paintings. Maps were an important part of this national envisioning of Korean society, combining a careful depiction of national territory with a form of national writing. The Korean word for map, *chido*, can be translated as "land picture." As pictures of the land, maps were systems of surveillance and control but were also vehicles for national representation, unification, and identity, important for envisioning, recording, and writing the nation. Early Korean cartography has a strong pictorial element. Ahn (2008) points to the many connections between early cartography and painting. Professional painters also made maps, and many early maps have a painterly quality, with painting techniques used in representing seas and rivers. From the fifteenth century to the eighteenth, elaborate, sinuous renderings of waves depicted the seas around Korea. Figure 8.1 is a very good example. Pictorial maps were common at the regional and local scales. There are three main forms: the panorama, which uses a bird's-eye aerial perspective (fig. 2.1); the blooming flower, in which mountains are shown as spreading out from a central area (fig. 2.2); and the closed flower, in which mountains tilt inward onto a central site (fig. 2.3). They display a spatial sensitivity and an aesthetic concern with the depiction of site and location.

NATIONAL MAPPING

Perhaps the most famous of the early Joseon kings was Sejong. His image is on one side of the 10,000 won note; the other side depicts scientific instruments to make the point about his enlightened rule. To rule effectively, the new regime needed information. In 1424 King Sejong ordered a survey of the nation. Questionnaires were sent to each governor and magistrate in 334 districts, requiring information on boundaries, population size, administrative history, distances to neighboring district centers, and an array of data on the human and physical geography of local areas. The survey created a new national geographical information system that was updated in 1469. In the 1470s another survey was launched that resulted in the 1481 *Complete Conspectus of the Territory of the Eastern Country [Korea]*. This survey was regularly supplemented and updated until the final edition of 1550. Note how the title *Eastern Country* defines Korea in relation to China.

In 1434 Sejong required local administrators to furnish "full and detailed maps." Maps of local areas assisted in administrative control by strengthening centralized geographic information and hence political power. Sejong's main cartographer was Chong Ch'ok (1390–1475), who produced a number of maps for Sejong as well as

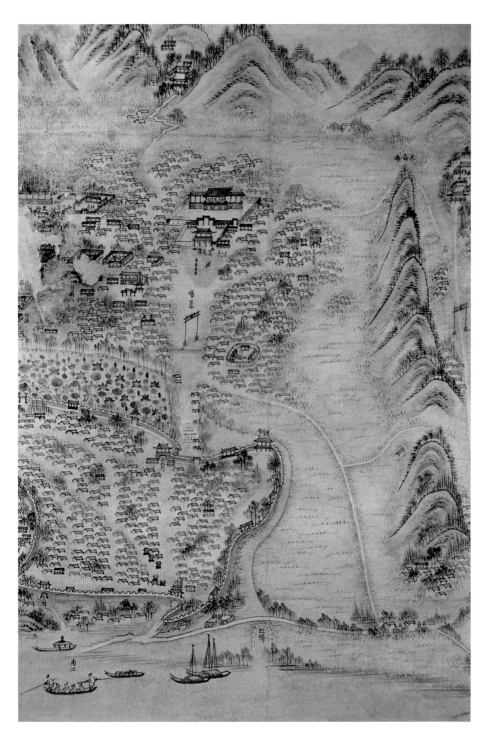

2.1 JINJU COUNTY, nineteenth century. (Kyujanggak Archives, Seoul National University.)

2.2 CHEOLONG CASTLE, eighteenth century. (Kyujanggak Archives, Seoul National University.)

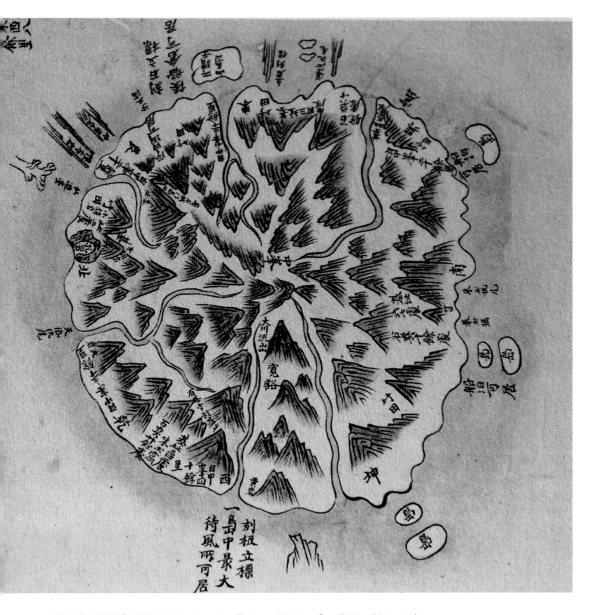

2.3 ULLUNG ISLAND, eighteenth century. (Kyujanggak Archives, Seoul National University.)

for subsequent rulers. Systematic mapping of provinces and districts thus began and was standardized to a common scale in the eighteenth century.

One important variant of official detailed maps was *kwanbangdo* maps, emphasizing terrain and military communications to assist with the country's defense. They were produced in a variety of forms, from long scrolls showing boundary regions to detailed maps of fortresses and coastal defenses. *Kwanbangdo* maps were often painted on screens or drawn on portable scrolls for use in the field by military officers and commanders.

Border areas were of special interest in the early Joseon period because of territorial expansion in the north. The northernmost extent of Silla control was just north of Pyongyang. The border was pushed farther north in the Goryeo period and even farther in the early Joseon, reaching its northern limit about 1441. This territorial addition, while nominally under Joseon control, was more of a frontier zone and difficult to manage and administer. The final northern border was fixed only in 1712, in an agreement with the Chinese. For the early Joseon, then, the northern border was of intense interest, since it was the liminal area of state control. Chong Ch'ok made a number of maps of two frontier zones, and numerous maps of these zones were also made by other cartographers including Yang Songji and Yi Sunsuk.

Sejong's son Sejo (r. 1455–68) carried on his father's cartographic concerns. While still a prince, but already the de facto ruler, he commissioned maps of the frontier zones as well as more general maps of Korea and of the separate provinces. In 1463 Sejo, now king, was presented with a map of the eight provinces. This map is very important because it laid the basis for depictions of the country from the middle of the fifteenth century to the middle of the seventeenth, termed Chong Ch'ok style. By putting the provinces in one text, the country was depicted as a coherent whole, an administratively linked national society—a persistent cartographic concern of the early Joseon. As early as 1400, Yi Hoe made a map of the eight provinces, and Chong Ch'ok and another cartographer, Yang Songji, made maps of these provinces for King Sejong as well as King Sejo. Both mapmakers were responsible for the 1463 map. The original is lost, but there are enough copies to give us some idea of the national representation at this time. Figure 2.4 shows a map of Korea from a printed atlas made about 1716–30, based on an earlier Chong Ch'ok map from about 1451. Provincial borders are in red, and the colored oblong symbols represent county seats. Seoul is shown as a walled city. Figure 2.5 is a detail from Cholla Province in the southwest of the country. Notice that the emphasis is on the administrative geography, the political hierarchy from Seoul to province to county seats.

The Chong Ch'ok style maps are recognizable by the depiction of the northern

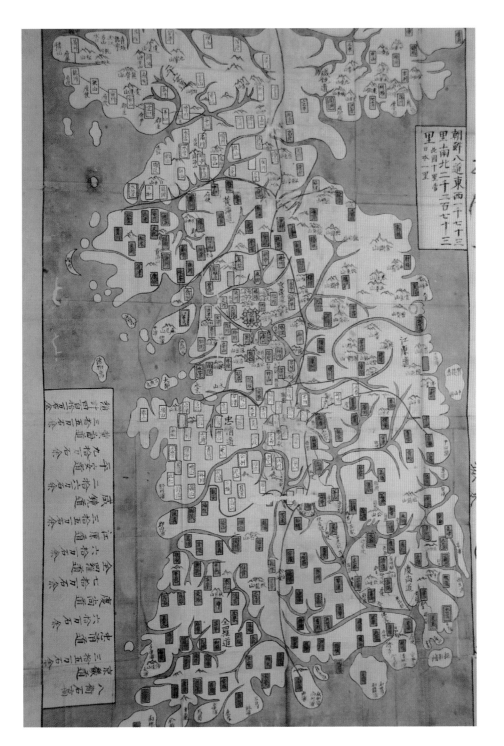

2.4 KOREA, 1730. (Courtesy of the Library of Congress, Washington, DC: G7900 145-.C4 Vault.)

2.5 DETAIL OF CHOLLA PROVINCE, 1730. (Courtesy of the Library of Congress, Washington, DC: G7900 145-.C4 Vault.)

2.6 NORTHERN REGION OF KOREA. (Courtesy of the Library of Congress, Washington, DC: G2305 D35.1782 Vault Shelf.)

frontier zone. Figure 2.6 is from a 1721 atlas in this style. Note how the border region is flattened, Manchuria is filled with sinuous rivers that simply fill up the space on the map with a hazy geography rather than depicting an accurate hydrography. Notice, however, the carefully delineated outline of the extinct volcano Baekdu (White Head Mountain), long regarded as a sacred site. Ledyard (1994, 292–94) suggests that the hazy depiction of this border area may be conscious attempt to mislead enemies who might gain access to the map. This national security consideration, embedded in the earliest maps, was copied by later mapmakers, allowing us to follow the progress of the Chong Ch'ok variant down through Korean cartographic history.

Sejo's main cartographer was Yang Songji, who left a list of maps he worked on or was familiar with up until his death in 1482 (Ledyard 1994, 345). The list of twenty maps reveals the new dynasty's concern with territorial integrity, with more than six maps of border provinces and frontier zones and ten maps of all or some provinces. One map was of coastal shipping routes, one was of Ming China, and one was of Japan. Official cartography involved mapping the frontier zones and demarcating administrative areas. This list of maps reveals the regime's concern with external security along the northern border, the internal order of provinces, and the coherence of a national state.

The Chong Ch'ok style was not the only cartographic representation of Korea in the early Joseon. There is another variant termed the sungnam style, based on a Ming Chinese text. I have already noted the creation of the *Complete Conspectus of the Territory of the Eastern Country [Korea]*. This gazetteer rigidly followed an earlier Chinese work, the *Comprehensive Gazetteer of the Great Ming*, completed in 1461. The maps associated with this Chinese text have a distinctive look: rivers and coastlines are gently curving rather than carefully delineated, mountains are shown by the briefest of brushstrokes, and the spaces of the map are dominated by cartouches containing the names of administrative districts. It is a political geography map drawn up by and for administrators, with little attention to other aspects of human and physical geography. In 1485 King Songjong called for the printed edition of the *Complete Conspectus* to follow the earlier Chinese model. Figure 2.7 is part of a large map of Kyongsang Province made according to this model about 1800. Physical features of mountains and rivers are diagrammatic rather than drawn to scale, and the typography of the round cartouches overwhelms the geographical message of the map. It is more typographic text than map. This style of provincial map was a popular and recurring feature of Korean cartography that continued down to the nineteenth century.

HYONGSEDO: SHAPES AND FORCES

While the administrative needs and ideological requirements of the Joseon dynasty dictated the subject of official mapmaking, many maps drew on a distinctly Korean tradition of *hyongsedo* that emphasized the vitality of the land. The tradition began much earlier but was systemized by Zen monks, such as Doseon (c. 827–98), who surveyed much of the country at the end of the united Silla period. *Hyongsedo* maps, a term that can be translated as "shapes and forces," have a forceful energy, owing much the rich tradition of geomancy, which sees the landscape as alive with energy or life force, known in Chinese as *chi*. A correct reading of the landscape to maximize positive energy was essential for the proper siting of graves, shrines, houses, temples, palaces, and other buildings. In this tradition landscape is not inert but is a vital living force.

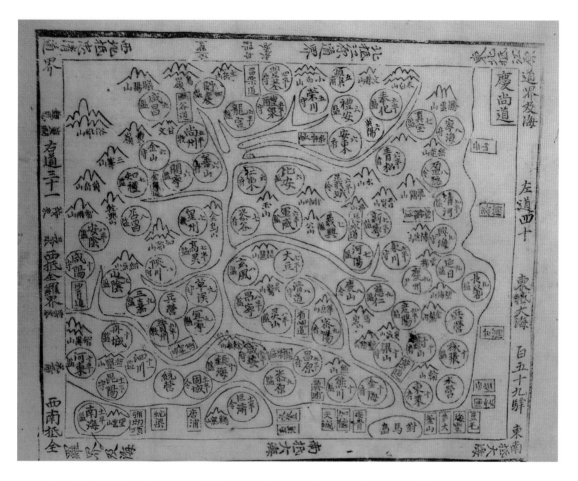

2.7   KYONGSANG, untitled scroll, c. 1800. (Courtesy of the Library of Congress, Washington, DC: G2330.C53 18—.)

The *hyongsedo* maps derive from three interlinked belief systems. The first is the shamanism brought from the Siberian forests by the earliest people to populate the peninsula. These early ideas included a pantheistic belief in the powerful spiritual energy in all living things. The second influence was the geomancy imported from China. Chinese writers wrote that the proper siting of things could maximize the positive energy in the landscape. Human action could tap into this vital energy with the proper placement of graves, buildings, and towns. The land was imagined as a living, breathing force with mountain chains as arteries and rivers as veins. A similar view held sway in much of premodern Europe. Here is Leonardo da Vinci describing a belief common in Renaissance Europe:

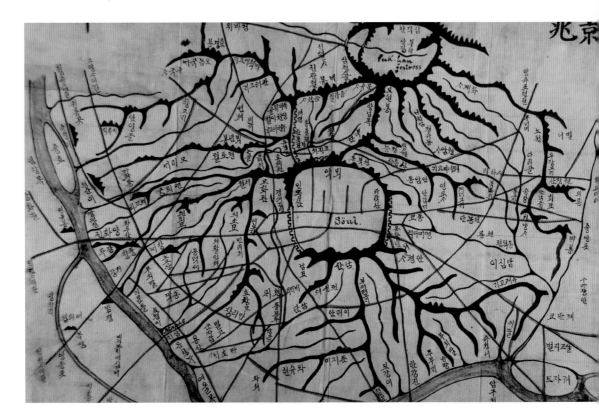

2.8 SEOUL, 1883–87. (American Geographical Society Library, University of Wisconsin-Milwaukee Libraries: 469-d S46 A- [between 1883 and 1887].)

While man has within him a pool of blood. . . . he breathes and contracts. So the body of the earth has its ocean, which also rises and falls every six hours with the breathing of the world. . . . the earth has a spirit of growth, that its flesh is the soil, its bones the arrangement and connection of the rocks of which the mountains are composed, its cartilage the tufa, and its blood the springs of water. (Richter 1952, 45–46)

Seeing the earth as a giant organism is a common belief system of premodern societies. In Korea as elsewhere, modernity did much to divest the physical environment of its wider symbolic meaning and deeper spiritual significance (Short 2000).

The third, more local, influence was the geomancy important in the pre-Joseon Goryeo dynasty, which visualized powerful forces coursing through the nation, linking sacred sites to political capitals. The oldest extant Korean map, the

2.9 GWANGJU COUNTY, 1872. (Kyujanggak Archives, Seoul National University.)

2.10 KOREA, 1730. (Courtesy of the Library of Congress Washington, DC: G2305 D35.1782 Vault Shelf.)

Gangnido (fig. 1.1), which derives from the knowledge of the Goryeo dynasty, depicts the physical geography of Korea in a geomantic way.

The shapes and forces maps were made at all scales. At the local scale they were used to site graves and temples. They were also used in establishing the new Joseon capital of Seoul. Figure 2.8 is a map of the capital region that looks more like a body's arteries and veins than a map of a city. Even at the county, provincial, and national scales the influence of shapes and forces can be discerned in the animate nature of this cartography. Figure 2.9 is a map of Gwangju County. Made much later in the Joseon reign, about 1872, this map highlights the enduring legacy of this type of cartographic representation. Figure 2.10 is a map of the nation taken from a later copy of a map in the Chong Ch'ok style. The rivers and mountains run though the body of the country giving life as well as form; they are circuits of energy, flows of power. Korean maps of this type have a palpable vitality.

The maps also have a distinctive design because they identify watersheds. They are a sophisticated hydrological rendering of the landscape.

3

EUROPE LOOKS EAST

Although Korea had a strong mapmaking tradition, from a European perspective Korea emerges slowly out of the confusion of a barely known Orient. The world-view depicted in figure 1.2 was the standard European perception in the fifteenth century. The map shows only half of the world, just 180 degrees of the 360 degree globe. East Asia is at the very edge of the known world, a space of locational uncertainty and geographical confusion.

The Europeans knew of East Asia through trade. The Silk Road, an interconnected series of trade routes linking East and West, had been a well-traveled highway since at least 200 BCE. The Silk Road not only was a route for commodity exchange, it was also a transmission belt for culture. Ideas and practices that moved along the road were modified and transformed in their diffusion across Eurasia. The central tenets of Buddhism, for example, moved east and north along the route from India to China and Korea. The road network crossed different cultures that allowed an innovative and dynamic mingling of artistic influence. Greco-Buddhist art developed as the classical forms of Greek culture were adopted into representations of the Buddha. The art percolated through to China and later into Korea in the sixth century.

The Silk Road was a route that could be followed without deep geographical knowledge. You put one foot in front of the other along a well-worn track. The European knowledge of Asia, especially East Asia, was hazy at best, and how it fitted into world geography was only guessed at. A major improvement in geographical understanding came through Venetian traders. Venice was a powerful commercial empire centered in the Mediterranean, with links through the Muslim world to East and Southeast Asia. Trade with the East provided access to precious commodities, especially spices and silks that could be sold at a premium in Europe's expanding markets. Two Venetian merchants, the brothers Niccolò and Maffeo Polo, established trading posts in Constantinople, then moved east to the Crimea, and then even farther east to Bukhara in present-day Uzbekistan. In 1264 they joined a diplomatic mission sent from that city to what is now Beijing, arriving after a two-year journey. They eventually returned bearing a letter from Kublai Khan to the pope. The two brothers, along with Niccolò's son Marco, then returned to China, where they spent the next seventeen years. The khan took a liking to Marco Polo and sent him on various official missions around the empire. On his return to Europe, Marco Polo wrote a book that became very popular. He described a large, populous island, Chipangu (Japan), to the east of China and imparted information on Korea he obtained from the Chinese. Christopher Columbus, among many others, read the book avidly.

The veracity of Polo's book is still not entirely assured. But even if he falsified his experiences, he could draw on the accounts of other travelers. The Italian Franciscan missionary Giovanni da Montecorvino (b. 1247) traveled to Persia, India, and China, and in the early fourteenth century another Italian Franciscan, Odorico da Pordenone (d. 1331), founded the first Christian mission in Beijing before returning to Venice. Polo's writings and his knowledge of China (he reputedly brought back a Chinese map of the world and a nautical chart) influenced European mapmakers and globemakers. The island of Chipangu (in various spellings) appears in numerous maps and globes such as the 1375 *Catalan Atlas*, a world map drawn by the Venetian monk Fra Mauro about 1450, and Martin Behaim's 1492 globe. The Mauro circular map of the world with south at the top, depicting a sea route around the tip of Africa, is one of the first European maps to show the islands of Japan and to suggest the peninsula of Korea. This map uses the full extent of medieval European cartography and the reports of Marco Polo but also relies heavily on Arab, Chinese, and Persian sources. Ibn Battuta, an Arab contemporary of Marco Polo, visited Central and Southeast Asia as well as China. There is also the possibility that the map owes some of its information to the mapmakers associated with the early voyages of Admiral Zheng He and his treasure fleet. From 1405 to 1433 Chinese emperors placed He in charge of seven naval expeditions that sailed throughout the Indian Ocean.

Although the Mauro map did depict the peninsula of Korea more accurately, at this early stage in the global production of space, it did not form the basis for subsequent geographical representation. Because of the lack of a singular global geographical knowledge, Korea was subsequently depicted in many configurations.

AN EXPANDING EUROPEAN VIEW

The western European powers, especially the English, Portuguese, Spanish, and Dutch, did not have as easy land access to oriental goods as the Venetians did; they had few connections in the Middle East, and their merchants had to traverse vast distances over routes easily and regularly cut off by intervening powers. They became seaborne empires that needed accurate maps to sail the roadless seas and trade across vast oceans. The rise of the western European powers looking for sea trade generated more accurate cartographic representations of the Orient.

Mapmaking in Europe went hand in hand with rising political power and efforts at overseas expansion. The center of mapmaking shifted from Italy first to Portugal then Spain, and then on to the Netherlands and later to England as commercial empires waxed and waned. Exploration and mapmaking were vital elements in European rivalries and attempts at commercial dominance.

The prevailing economic ideology of the time was mercantilism, seeing foreign trade as the chief method of increasing national wealth. The mercantilists believed the world's wealth was like a giant cake, fixed in size: any increase for one nation came at the expense of the others. The most favorable conditions for trade thus depended on monopoly. Successful commerce implied a commercial empire where prices could be fixed, markets protected, and competitors kept out. The overseas expansion of European influence was a commercial undertaking driven by mercantile capitalism. Empires grew as nations pursued dominance over trade and territory in Africa, Asia, and the New World.

The Portuguese sailed south and east. In 1443 Prince Henry secured a monopoly over trade in West Africa and began a policy of active exploration to chart the African coastline. In 1456 Alvise Cadamosto sailed up the Gambia River, and soon the Portuguese were shipping thousands of slaves back to Portugal. In 1462 Pedro de Sintra reached present-day Sierra Leone. Maps of the coastline then appeared in various portolan maps (sea charts), including those by the Venetian Grazioso Benincasa, drawn in 1462 and 1468 from Portuguese sources. The linguistic legacy of these early Portuguese voyages remains: Lagos in Nigeria is named after a town in southern Portugal.

Through the latter half of the fifteenth century the Portuguese continued to move farther down the African coast. Portuguese ship captains pushed the limits of geographic knowledge in successive voyages. Fernão Gomez passed the equator in

1473. Nine years later Diogo Cão passed the mouth of the Congo River. The Portuguese found out that India could be reached by sea, and in 1487 Bartolomeu Dias sailed all the way around the Cape of Good Hope (named by Portugal's King John II) to the Indian Ocean. The Portuguese captains were encouraged to keep careful records and accurate logs, but since this was very valuable commercial information, the results were kept secret in official maps with restricted access. But then as now, valuable commercial information rarely stayed secret for long, and the results of the voyages were soon recorded in sea charts and eventually appeared in printed maps showing a southern ocean that allowed passage from the Atlantic to the Indian Ocean. The African discoveries were documented in various Portuguese maps including a portolan map made about 1490 by Pedro Reinel.

After consolidating power against the Moors, the Spanish Crown also looked overseas. The Spanish funded the voyages of the Genoan Christopher Columbus. His initial plan was to find a sea route to Asia and gain access to the riches and spices of the East. He sailed on August 3, 1492, with three ships. On the night of October 11, they sighted land. He first landed on a small island in the Bahamas and, thinking he had reached the East Indies, called the local Taino tribe "Indians." But as we now all know, he had encountered a New World, a landmass that occupied part of the 180 degrees missing in the Ptolemaic world maps.

The growing competition between Portugal and Spain for overseas territory finally led to an agreement to divide up the world. In 1494 the Treaty of Tordesillas signed between Portugal and Spain, brokered by the pope, effectively divided the newly discovered world in two. Territory east of a line drawn along one half of the earth roughly between forty-two and forty-six degrees west longitude (the line was never accurately measured) was ceded to Portugal. In the wake of this decision the Portuguese king sent Vasco da Gama around the southern tip to India to access the land now opened up to Portugal by the treaty. Four ships sailed on July 8, 1497, reaching the port of Calicut in India on May 29, 1498. From their base in Goa, in India, other Portuguese explorers pushed even farther east. In 1511 Alfonso de Albuquerque, with twelve hundred men in a score of ships, captured Malacca in present-day Malaysia. A year later the Portuguese established a base in the Spice Islands. And in 1513 Jorge Álvares landed in Guandong, on an island at the mouth of the Pearl River. The Portuguese authorities in Goa send emissaries to inaugurate trade with China. The Chinese granted Macao, south of Canton, to the Portuguese, who then entered the trade between Japan and China and the wider regional trading circuits. In 1542 Fernão Mendes Pinto landed in Japan. The area was also described in numerous travel books. Pinto, for example, in his *Peregrinação* (Pilgrimage), published in 1614, described his travels in the Far East, China, and Japan undertaken between 1537 and 1558.

One of the earliest surviving maps to show the Portuguese discoveries in the East is the Cantino world map of 1502. Alberto Cantino was a spy for the Duke of Ferrara, sent to Lisbon specifically to learn Portugal's geographical discoveries. Cantino copied this map from a secret official map kept in the Casa da Índia, the main repository for new geographical information. The pirated map shows an open Indian Ocean, records India, and details the coastal outlines of Southeast Asia. The map shows the extent of Portuguese reach into Asia at the beginning of the sixteenth century. By the middle of the century the Portuguese had explored and mapped their way east to the Spice Islands. They used local maps and charts. Early sixteenth-century documents note the use of a Javanese world map that recorded navigation routes used by the Chinese and other peoples (Olshin 1996). A portolan chart made by Pedro Fernández in 1545 stands at the limit of European knowledge before the more detailed exploration of the region. The Spice Islands of Southeast Asia are shown disproportionately large, a reflection of their economic importance and perhaps the use of local charts. The Fernández map is a record of the knowledge that seaborne western European powers had of the region around the middle of the sixteenth century. China is known, but Korea remains unknown.

KOREA BECOMES KNOWN

The exploration of the Far East was undertaken by European powers eager to establish trade links and monopolize the supply of precious goods and commodities, especially spices such as nutmeg and cloves. The economic rationale is clear when the profits from the trade are calculated. Nutmeg bought directly in the Spice Islands of Banda, for example, cost English traders just one penny for ten pounds by weight. In London the same amount was sold for two pounds, ten shillings, a markup of 60,000 percent (Milton 1999, 6). The lure of these windfall profits made the islands and indeed the whole region a site of bitter conflict between the Portuguese, the Dutch, and the English.

The lure of commercial riches attracted other seaborne European powers eager to break the early Portuguese monopoly. The Treaty of Tordesillas restricted the Spanish to west of about forty-two degrees west longitude. But on the far side of the world the demarcation was not so clear. If the treaty line extended all the way around the globe—the line would roughly bisect Papua New Guinea and Japan—then the Spanish had legitimate entry to the Far East. They found a willing captain in Ferdinand Magellan, a Portuguese with extensive sailing experience. In 1505 he sailed to India with Francisco de Almeida (Bergreen 2003). Magellan's proposal to sail west with an expedition was accepted by the Spaniards, in search of glory, gold, and geopolitical advantage over their rivals. Magellan set out in 1519, with three ships and 237 men. He sailed across the Atlantic, around the southern tip of South America—the narrow strait is now called the Strait of Magellan in his honor—and crossed the vast Pacific Ocean. On March 16, 1521, he was the first European to reach the Philippines. Although the islands were technically within the Portuguese sphere, they were on the other side of the world. And possession on the ground trumped legal agreements in a distant Europe.

In 1566 the Spanish formally claimed the Philippines and made Manila an entrepôt for trade with China and Japan. Spanish galleons subsequently made regular Pacific crossings carrying silver from the mines in the New World. Spanish silver became a major medium for purchasing Chinese goods.

The Dutch also became major players in the region. The Dutch Republic was established in 1579 and soon set about achieving wealth and power by creating a worldwide trading system. Overseas trade and commerce were the lifeblood of the new republic. By 1600 the Dutch had almost ten thousand ships sailing around the coasts of Europe and across the oceans to trade in grain, tobacco, barley, herring, timber, sugar, and spices. The tentacles of Dutch trade quickly stretched from their Baltic origins to the Mediterranean, across the Atlantic, south to Africa, and across the Indian Ocean to India, Southeast Asia, and the Far East. At home, a vigorous merchant community was successful in establishing a commercial society. The Bank of Amsterdam, founded in 1609, soon achieved world prominence. At this time a wide variety of coins and currency were in circulation. The bank took them all, assayed the gold and silver content, and allowed depositors to withdraw the equivalent in gold florins minted by the Bank of Amsterdam. The bank became a depository of huge holdings and a central exchange of global financing. This newfound wealth was filtered through the frugal ideology of a Calvinist theology. The dilemma of how to be wealthy and moral at the same time helped give shape and substance to a distinctively Dutch culture (Schama 1987). The Dutch Republic, with Amsterdam its center, became a shipping center, commodity market, and capital market for the world economy.

In 1595 a Dutch fleet sailed from Amsterdam to the Spice Islands in open defiance of the Portuguese monopoly on the trade, making a direct assault on the primacy of Lisbon as the only European port for importing East Indian goods. While the Spanish and Portuguese were experiencing the beginnings of imperial decline, the Dutch were on an upward course. Hailing from a small Protestant country, driven more by commerce than by religious fervor or imperial pretensions, the Dutch were consummate traders. In 1598 a Dutch fleet followed Magellan's route to the Moluccas. One of the ships eventually sailed to Japan, and the English pilot William Adams stayed in the country acting as a go-between for the later European traders and the shogunate.

The commercial imperialism of the Dutch, as well as of the English, was a private matter rather than a concern of the state. In 1600 the English East India Company was established by a group of merchants eager to break into the lucrative trade with the Far East. By 1621 the company had detailed charts and maps of the Spice Islands and neighboring regions (Tyacke 2008). The English Crown and the States General, the ruling body of the Dutch Republic, did have a regulatory role, however. They could grant commercial privileges and bestow a monopoly of trade on groups of merchants. Although these pronouncements had little power over foreign merchants, they could stimulate the commercial undertakings of their own merchants and control competition between them. The establishment of the Dutch East India Company in 1602 was formal recognition of the Dutch involvement in the Far East. The company amalgamated all the Dutch trading companies and strengthened the hand of Dutch military commanders in the region. The Dutch traded through Taiwan, the Chinese city of Canton, and the Japanese city of Nagasaki. They become the sole European traders when Japan expelled the Spanish and Portuguese in 1641.

European influence and control in East Asia eventually centered in the three cities of Canton, Nagasaki, and Batavia, Indonesia (Blusse 2008). By the middle of the seventeenth century the Dutch, Portuguese, and Spanish had bases in China including Taiwan and Japan. They traded through the cities of Canton and Nagasaki and also from their own bases: the Portuguese from Macao and the Japanese island of Hirado, the Dutch from Taiwan.

There were also Christian monks, priests, and missionaries in the region. The Jesuits especially came to the Far East. The Portuguese Jesuit Luís Fróis lived in Japan for thirty years from 1563. His writings provide illuminating descriptions of his perception of contemporary Japanese culture and society. He describes, for example, thousands of Korean prisoners of war in the city of Nagasaki after the Japanese invaded Korea in the 1590s. As literate and educated men, the Jesuits become important recorders of the local scene, and many served as advisers to the ruling powers. Matteo Ricci entered China from Macao in 1582 and became an adviser to the court, and other Jesuits served as scientific advisers to the Chinese emperors for the next fifty years. Their involvement in official mapping exercises led to a distinct Sino-Jesuit cartographic tradition. I shall be speaking more of such figures and their pivotal role in mapmaking.

While China and later Japan begin to be familiar to the European merchants and missionaries, Korea was still mysterious. Farther north and thus not offering spices, the Korean ruling elite were not at all keen to make strong connections with the outside world in general, let alone European traders. The Europeans nevertheless learned about the country through a variety of sources. The Japanese and the Chinese had some very detailed information. Korea was also known to local sailors and fishermen and represented in local sailing charts and maps that eventually made their way into the European cartographic repertoire. The Europeans picked up information as they traveled around the region and lived in the towns and cities of China and Japan. The Christian missionaries were also a major source. The first European recorded as visiting Korea was Gregorio de Céspedes, a Spanish Jesuit priest who accompanied an invading Japanese army in 1593. But there were no written accounts of firsthand visits by Europeans until the middle of the seventeenth century.

In summary, in 1500 European-recorded knowledge of Korea consisted of little more than a few sentences written by Marco Polo. By 1600 there was more material, largely from Jesuit missionaries in Japan, who mention Korea in letters sent back to Europe (Cheong and Kihan 2000). The Jesuits spoke with Japanese and Chinese officials as well as with merchants and sailors and perhaps crossed paths with Korean envoys. The Jesuit letters were compiled in the *Annual Letters of Japan*, first published in 1593. Several were also reprinted in the very influential second edition of Richard Hakluyt's *Principal Navigations*, published in 1599–1600. In 1601 Luis de Guzmán compiled the *Historia de las Misiones*, which summarizes the missionary works in the West Indies and Far East and contains eighty pages on Korea. Guzmán correctly identifies Korea as a peninsula.

Korea also appears in early Dutch writings. Jan Huyghen van Linschoten (b. 1563) who worked for a while as an assistant to the archbishop of Goa in India, published his *Travel Notes* in 1595 and mentions, "Cooray" as a place on which he has "good, comprehensive and true information." He also referred to Korea as Chosun. His book was translated into several languages, including English, Latin, German, and French.

There were some early visits. The first recorded sighting took place in October 1578, when a merchant ship sailing from Macao to Japan was driven off course by a storm. One of those on the ill-fated ship, Father Antonio Prenestino, an Italian priest, later wrote home, "Korea, barbaric and inhospitable people, desires to have dealings with no other people" (De Medina 1991, 39).

KOREA IN EUROPEAN MAPS

We can take 1550–1650 as an approximate era for increased European knowledge of the Far East, owing to trading links and various reports and accounts, but also as a time of limited knowledge of Korea. In this period of cartographic confusion, when a global space was in embryonic form and still full of ambiguities and inconsistencies, Korea appears in a number of ways. I identify three distinct representations: Korea unknown, Korea as an island, and Korea as a peninsula.

Korea Unknown

In many popular European geographical accounts, Korea remains unknown. Figure 3.1, for example, shows the depiction of East Asia by Sebastian Münster. His major work, *Cosmography,* was first published in Geneva in 1544. The first edition of this massive work had 659 pages, with 520 woodcut maps and illustrations. By the 1550 version, the work had reached gargantuan proportions, with 1,233 pages and 910 woodcuts. It was published in all the major European languages as well as in Latin. Thirty-six complete editions were published between 1544 and 1628. For much of the sixteenth century *Cosmography* was the great educational book of its era, the single most important source of geographical, historical, and scientific knowledge. An eclectic collection of material, part myth, part fact, Münster's work encapsulates both new knowledge and older conceptions of the world. Figure 3.1 shows that *Cosmography* does give prominence to the Spice Islands of the East Indies, but Korea remains hidden and unknown. There is a scatter of tiny islands shown off the coast of China, but nothing to indicate peninsular Korea. The most popular European geographical text of the sixteenth century had no depiction of Korea.

A more self-consciously modern geographical work is the first modern atlas by Abraham Ortelius, the *Theatrum Orbis Terrarum,* first published in 1570, an important document in the global production of space. There had been collections of maps in one volume before Ortelius, but he set a standard by which subsequent collections would be judged. He was born, raised, lived, and died in Antwerp, a thriving, dynamic merchant city, and always referred to himself as a citizen of Antwerp. In 1560 the city had a population of about 100,000, with almost 600 foreign merchants. It was a crucible of European mercantile capitalism, full of rich merchants with plenty of disposable income. The *Theatrum* was begun about 1566–67, completed in 1569, and sold to the public in 1570. Ortelius sought out the most accurate current information and collected maps from a wide range of cartographers. He listed the names of all the cartographers whose work he had used: 87 were mentioned in the first edition of 1570 and 182 in the 1603 edition. The *Theatrum* was a huge success; between 1570 and 1598, it sold 2,200 copies. It was published first in Latin and later in Dutch, German, French, Spanish, Italian, and English. Additions were made to later editions. The 1570 edition had 50 sheets, and the Italian edition of 1612 had 129. The book was being reprinted as late as 1724, and almost 7,300 copies in eighty-nine editions were eventually printed. The *Theatrum* provided Ortelius

3.1 MÜNSTER'S ASIA, 1540. (Sebastian Münster, *La cosmographie universelle* [Basle: Henry Pierre, 1551]. Courtesy of the Library of Congress, Washington, DC.)

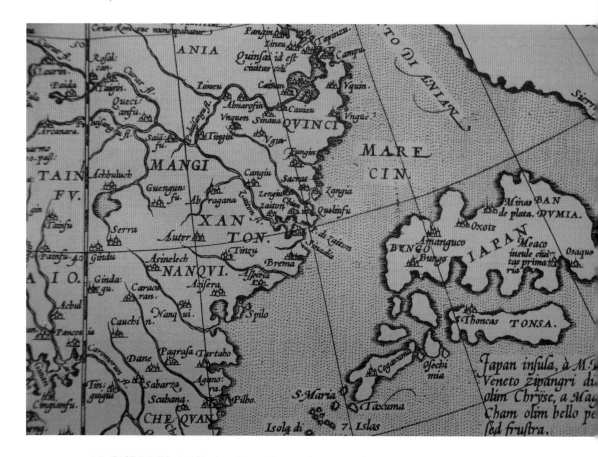

3.2 EAST INDIES. 1570. (Abraham Ortelius, *Theatrum Orbis Terrarum* [Antwerp: Gilles Coppens de Diest, 1570].
Courtesy of the Library of Congress, Washington, DC.)

a good living. He lived a burgher's happy life, moving into ever larger, more expensive houses in Antwerp.

Ortelius covers the Far East in two maps, one of Tartary and one of the East Indies. Part of this second map is shown in figure 3.2. Notice the spatial grid composed of lines of longitude and latitude, suggesting greater geographical sophistication than in Münster's general description. But notice that while Japan has been added to the region, Korea remains unknown. The gridded map has elements of the new geographic knowledge, but its failure to depict Korea still reflects a Ptolemaic haziness. In this textual space of new and old geographical knowledge, Korea still remains hidden.

The emergence of Korea for Europeans is part of the creation of a global space of knowledge that is such a defining feature of early modernity (Parker 2010). The incorporation of Korea into European geographical knowledge systems—and, as I shall show later, the diffusion of the European worldview that affects Korean cartographic practice—was part of the understanding and indeed creation of a global space and its representation. It was not a straight path.

Korea as an Island   Because there was such limited European information on Korea, much of it based on secondhand information, hearsay, and rumor, its continuing cartographic depiction included many inconsistencies. Figure 3.3, for example, shows a map that accompanied Jan Huyghen van Linschoten's *Travel Accounts*. Linschoten was a merchant who lived in Spain and became an assistant to the archbishop of Goa. He obtained secret Portuguese sailing directions to the Far East. On his return to the Netherlands he published two influential books, *Travel Accounts of Portuguese Navigation on the Orient* (1595) and *Travel Accounts* (1596). It was in the 1596 book that he made this observation: "A little above Japan, on 34 and 35 degrees, not far from the coast of China, is another big island, called Insula de Core, from which until now, there is no certainty concerning size, people, nor what trade there is" (cited in Savenije 2009).

Figure 3.3 is a map from the 1596 book. It draws from both Portuguese and Spanish maps, especially in its depiction of the Philippines. Notice how much detailed information there is for many of the islands of Southeast Asia. But also notice that while the coastlines of Japan and China are annotated with place-names, Korea is represented as a small, almost blank island. Its representation as an island reflects Europeans' lack of solid and widely agreed on information about the kingdom as late as the end of the seventeenth century.

Figure 3.4 shows Korea as an elongated island. This was a common feature of European cartographic representation around 1600. The map appears in the 1596 edition of the *Theatrum Orbis Terrarum* and signifies a greater geographical knowledge than the one from the 1570 edition shown in figure 3.2. In little less than thirty years, then, Korea emerges from nothing to cartographic representation. The global production of space proceeded less as a straight line than as a series of sharp jumps when new knowledge was quickly transcribed into an increasingly shared map of the world.

The Portuguese Jesuit and mathematician Luís Teixeira, who was cosmographer to the Spanish king, made the map depicted in figure 3.4, probably based on Jesuit sources in Japan. Notice that Korea is no longer a small circular island but has become elongated, closer to its peninsular reality. This depiction reflects the

3.3 EAST ASIA, 1596. (Jan Huyghen van Linschoten, *Itinerario* [Amsterdam: Everhardt Coppenchurch, 1596]. Courtesy of the Library of Congress, Washington, DC.)

increasing but still limited geographical knowledge of the country available to Europeans. This map variant was very popular, especially given the significance of the Ortelius atlas, and it circulated in numerous forms throughout the seventeenth century. It continued to appear in popular atlases as late as the eighteenth century.

There is also a cartographic variant that shows Korea as an island, yet so close to the mainland that it is almost a peninsula. This variant combines the more accurate peninsular form with an inaccurate island depiction. It appears in a variety of texts. The map of the world shown in Hakluyt's 1599 *Principal Navigations,* drawn by Edward Wright, depicts Korea so close to the coastline that it almost merges into the continental landmass. This same variant is repeated in the Mercator-Hondius-Janson atlas of 1636.

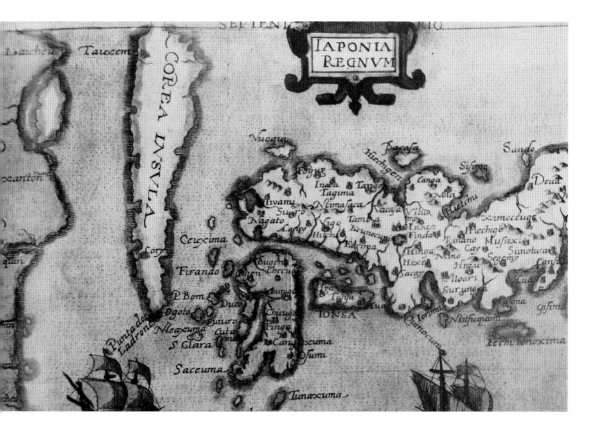

3.4 JAPAN AND KOREA, 1596. (Abraham Ortelius, *Theatrum Orbis Terrarum* [Antwerp: Plantiniana, 1596]. Courtesy of the Library of Congress, Washington, DC.)

Korea as a Peninsula   Korea was also correctly depicted as a peninsula. Diogo Homem, a Portuguese cartographer who worked in London and Venice, produced a number of maps and atlases. His manuscript map of 1588 shows Korea as a peninsula. João Teixeira, the son of Luís Teixeira, was also a map- and chartmaker who served for a time as the official cartographer and cosmographer to the Iberian authorities and thus was privy to both Spanish and Portuguese discoveries and maps of East Asia. His chart of the North Pacific done about 1630 correctly identifies Korea as a peninsula.

If Ortelius's 1570 atlas was *the* atlas of the sixteenth century, then Joan Blaeu's *Grand Atlas* of 1662 was *the* atlas of the seventeenth century. Willem Blaeu (1571–1638) was a mathematician, astronomer, and instrument maker. About 1605 he established a printing press near the center of Amsterdam, where he published various works on navigation, astronomy, and theology. He worked with his son Joan (1596–1673), and together they published a series of atlases. The first, issued in 1630, consisted of 60 plates, some original, some copied. In 1631 an expanded version was printed. And in 1635 a greatly expanded atlas was produced consisting of 208 maps in two volumes titled *Novus Atlas*, with an alternative title of *Theatrum*. This atlas was enormously successful, and the Blaeus built on their success, expanding their business. When Willem Blaeu died in 1638, Joan carried on the family business and also assumed his father's appointment as official cartographer to the East India Company. The *Novus Atlas* was expanded to three volumes in 1640, continually enlarged in successive printings until six volumes were produced in 1655. In 1662–65 the six-hundred-map *Atlas Major* (sometimes referred to as the *Grand Atlas*) was produced, marking the magnificent apex of seventeenth-century Dutch cartography. It was the most expensive book of the seventeenth century and remains the largest atlas ever produced. Volume 9 of the 1662 Dutch edition contains 28 maps of Asia, China, and Japan. Korea is shown as an island in one of the general maps of East Asia (see fig. 3.5). This map is derived from an even earlier atlas published in 1655 by Blaeu, the *Novus Atlas Sinensis,* which contains maps of the Chinese empire and Japan compiled by the Jesuit Martino Martini (1614–61). Martini was born in Italy, and after entering the Society of Jesus he embarked for the East Indies in 1640 with twenty-two other missionaries. He arrived in China in 1643 and set about gathering as much geographical information as he could. He has access to Chinese maps such as the *Guang yu to* (Enlarged Terrestrial Atlas) made by Luo Hongxian about 1541, first published in 1555 and republished five times up to 1799. Martini returned to Europe in 1658 and made his way to Rome by way of Amsterdam, where he met Blaeu and assisted with the 1655 *Novus Atlas Sinensis,* contributing seventeen maps that include East Asia, China, all the provinces of China, and Japan. It was the first European mapping of the Chinese empire and provided the first European maps of the Chinese provinces. There is no specific map of Korea; it appears only in the map of the Chinese empire and on the map of Japan. However, since that map draws on Chinese sources, Korea is correctly depicted as a peninsula. But there are few details. For China and Japan there is much more information. In Japan, for example, town symbols with crosses signify Catholic missions. While the external outline of Korea is now more accurate, internal details are sparse. The peninsula is located but barely understood.

The three views of Korea unknown, Korea as an island, and Korea as a peninsula exist together throughout much of the sixteenth and seventeenth centuries. As an example of the confusion of the time, Blaeu's *Grand Atlas* contains depictions of Korea both as an elongated island and also as a peninsula. In its long, slow journey

3.5 KOREA, 1662. (Joan Blaeu, *Grand Atlas* [Amsterdam: Blaeu, 1662]. Courtesy of the Library of Congress, Washington, DC.)

from *incognito* to *cognito*, the country is represented in a variety of ways. The most accurate depiction of Korea as peninsula emerges from the cartographic encounter between Chinese and European mapmakers. A full scientific geographical understanding of Korea by Europeans occurs only at the end of the seventeenth century and results from complex cartographic encounters. The global production of space, as the most accurate early European depiction of Korea attests, was in effect less a European invention than the result of many varied intercultural transactions.

PART

2

CARTOGRAPHIC ENCOUNTERS

4 JOSEON AND ITS NEIGHBORS

National cartographies, whether French cartography, British imperial cartography, or in this case Korean cartography, on closer inspection reflect and embody cross-cultural movements and transnational influences. A distinctively Korean cartography emerges from cartographic encounters and cultural exchanges with other nations and centers of representation. The earliest influence was China, followed by Japan and later Europe, especially as mediated through China. In this chapter I will look at each of the influences separately, although in reality they form a complex and dynamic interaction.

CHINA

Korea drew much from China, including writing, language, religions, and political philosophies. The Joseon kingdom held a Sinocentric view of the world and saw itself as an eastern neighbor and vassal state of what it considered the world's greatest power and the center of global civilization. Many Korean maps and atlases use the term *Dong* (East or Eastern) to describe the country. Kim Jeong-ho's great work can be translated as *Map of the Great Eastern Nation*. Korea is eastern only if you assume the center is in China. Japan, in contrast, did not identify itself with any country, describing itself as the land of the rising sun.

Chinese atlases were regularly reproduced and copied in Joseon cartography, and in atlases China often came first after a map of the world. Often a China of the past was depicted, since the Joseon rulers saw themselves as heirs to the Ming dynasty. A map of China in a Korean scroll atlas made about 1800 shows the boundaries of a Ming China long after the dynasty had fallen to the Qing in 1644.

The Joseon drew on a rich and long Chinese cartographic tradition. One of the earliest maps, discovered in a tomb in Hebei in China, dates from about 400 BCE. Maps inked on wooden boards have been dated as early as 200 BCE. Maps were also made on paper, wood, silk, and stone. They were sophisticated, drawn to scale and using abstract symbols. The scholar Pei Xiu (224–71) was responsible for many of the principles of Chinese mapmaking. He identified scale, location, distance, elevation, and gradient as important considerations when making maps. Pei Xiu emphasized the need for careful measurement and attention to geographic detail in mapmaking. A later mapmaker, Jai Dan (730–805), is credited with using a grid in constructing maps, and subsequent Chinese maps until about 1600 are often characterized by grids.

During the Tang (618–907) and Song (960–1279) dynasties, local governments regularly drew up maps every three to five years and submitted them to the central authorities. Since there were almost three hundred local governments, many maps were produced over the years, and a sophisticated cartography was created.

Another important figure in Chinese cartography is Zhu Siben (1273–1335). In the wake of the Mongol unification of Asia, a new shared geographic knowledge now included Muslim, Persian, and Arab geographical works and maps. About 1315 he drew from this rich source to produce a large map of China that was to form the basis of cartographic representation of the country for centuries. He also produced maps of the provinces, as well as surrounding areas including Korea, and an atlas. The atlas was revised by Lo Hung-hsien (1504–64) and printed about 1555 as *Guang yutu* (Enlarged Terrestrial Atlas); it continued to appear until the end of the eighteenth century. Between 1405 and 1433, the Chinese maritime explorer Zheng He led seven expeditions to the Pacific and Indian Oceans, greatly expanding the empire's geographical knowledge. The *Guang yutu* incorporates his voyages. This atlas contains a distorted map of Korea displayed on Zhu Siben's distinctive grid. The map of Korea was the result of close ties between Korea and China at that time. Korean princes lived in Beijing and held court. Maps were no doubt part of the cultural exchange, and the map of Korea in Zhu Siben's atlas was perhaps based on a Korean map.

There was a complex cartographic interaction between China and Korea over many centuries. As early as the tenth century Lu To-sun, an ambassador to Korea, brought back maps of Korea. Later, in the period 1068–77, the Chinese scholar Shen

Kua notes, "ambassadors came from Korea bringing tribute. In every Chinese city or provincial capital which they passed through they asked for local maps, and these were made and given to them. Mountains and rivers, roads, escarpments and defiles, nothing was omitted" (quoted in Needham 1959, 549).

There was a multifaceted trade in maps and cartographic information between the two countries, sometimes open, often illicit. Many Chinese maps made their way to Korea. A map made in 1330, *Map of the Diffusion of Instruction*, and another made about 1380, *Map of the Territories of the One World*, were taken to Korea about 1399 by the Korean ambassador to China, Chin Shih-heng. In Korea they were combined in 1402 into *Map of the One World and the Capitals of the Countries in Successive Ages*. This Korean map, from Chinese originals, which also uses the earlier work of Zhu Siben, is the well-known 1402 *Gangnido* map (see fig. 1.1). The *cheonhado* world maps—more on them later—also draw from Chinese sources. Although a distinctly Korean map, the *cheonhado* derives its form from Central Asian and Chinese Buddhist beliefs and from maps and mandalas from earlier centuries.

The Chinese central authorities produced maps for military, political, and administrative purposes. Careful mapping and cartographic knowledge were seen as important elements of political power and military intelligence. Korea appeared in many Chinese imperial maps of both the Ming (1368–1644) and Qing (1644–1911) periods. Figure 4.1, for example, is a map of China that includes Korea; it is a 1720 revised copy of a Ming map based in part on an earlier map made by Zhu Siben in the fourteenth century. Figure 4.2 shows how Korea was depicted in an early Qing map. This detail is taken from a woodblock print made about 1800 that is a copy of an earlier map but with some updates on river courses. Both maps reflect the Chinese imperial view that Korea was part of the Chinese empire, a vassal state on the far eastern edge but still belonging to the wider empire. In many Chinese imperial maps, only the western coastline of Korea is depicted. Japan rarely makes it onto the page.

The arrival of the Jesuit missionary Matteo Ricci marked another significant development in Chinese cartography. Matteo Ricci (1552–1610) was born in Macerata, Italy, on October 6, 1552. He joined the Jesuit order when he was only nineteen, and eleven years later was posted to the mission station in the Portuguese province of Macao. He eventually gained permission to enter China in 1582, and by 1601 he had reached the imperial capital of Beijing. A zealous missionary eager to make converts to Catholicism, he was rumored to have converted two thousand high-ranking Chinese officials. Ricci was able to exert such an influence because he was sensitive to local traditions and respectful of Chinese culture. He took the trouble to learn Chinese, and he adopted some of the clothes and language of a traditional Chinese scholar. A transcultural intellectual, Ricci was part of a larger

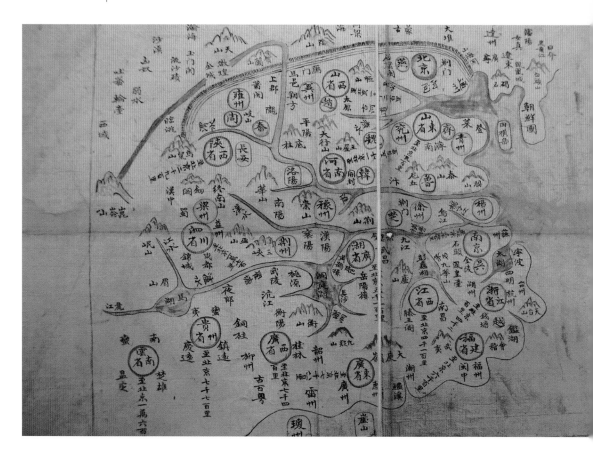

4.1   MING CHINA, 1720. (Courtesy of the Library of Congress, Washington, DC: G2305 D35. 1782 Vault Shelf.)

group of Christian scholars, including Michael Boym, Jean-Baptiste Régis, Ferdinand Verbiest, and others, who played an important role in mapping the Chinese empire. Ricci and his colleagues were transmission points between Chinese and Western cartography. They demonstrated to the Chinese the principles of latitude and longitude and introduced them to a less Sinocentric view of the world. And in turn they were introduced to the sophisticated range of Chinese cartography. The resultant Sino-Jesuit maps made their way from China into Korea. The Jesuits also introduced China and the Far East to the West by providing Western mapmakers with a more accurate picture of Asia.

A distinct form of Sino-Jesuit cartography emerges in the seventeenth century that involves both European Jesuits and local Chinese intellectuals. Ricci produced several world maps, including a 1584 woodblock print, a 1596 map carved on stone that was revised in 1600, and a 1602 map that is an enlarged and revised version of

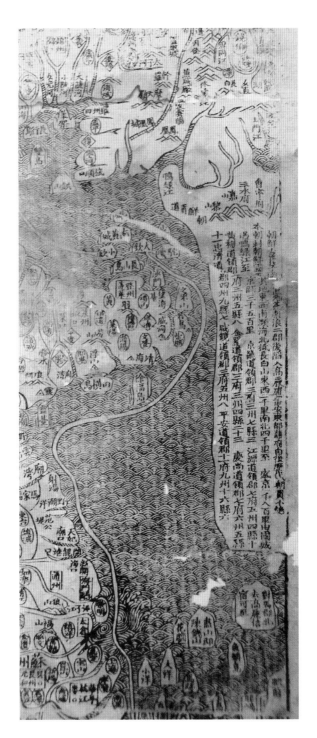

4.2 QING IMPERIAL MAP, 1800. (Courtesy of the Library of Congress, Washington, DC: G7820 1800. M3 Vault Oversize.)

the 1584 map. In his first world map of 1584, Korea does not appear. In the 1602 map it is shown as an oblong peninsula. Ricci drew the map in Beijing in 1602 with the help of a Chinese colleague, Li Zhizao, and in 1603 this Sino-Jesuit collaboration made its way into Korea to give a very different view of the world than the *Gangnido* or other Korean "world" maps (Bae 2008).

The confusion surrounding Korea in Ricci's early world maps may reflect the country's peripheral importance to the Chinese government and Chinese mapmakers. Korea was part of the greater Chinese empire but largely cut off from regular and heavy traffic. Later Sino-Jesuit maps had a more accurate depiction of Korea, such as Michael Boym's 1670 map of China or Ferdinand Verbiest's world map made for the Qing emperor in 1674.

A much more accurate Chinese representation of Korea comes only in the early eighteenth century as part of a broader imperial mapping project. In 1708 the Manchu emperor decided to map his entire extended empire, including Manchuria and Korea. He used Jesuit cartographers Erhernberg Xavier Fridelli, Pierre Jartoux, and Jean-Baptiste Régis. They were part of a group of Jesuits sent by Louis XIV of France in 1687, trained in the latest methods of survey and mapmaking. From 1708 to 1717 the European Jesuits, with help from local scholars, officials, and functionaries, surveyed the vast territory and mapped the empire. The Jesuits were not allowed to enter Korea, but they sent a Jesuit-trained Chinese surveyor to Seoul, accompanied by a senior Manchu envoy. The two were closely watched as they made careful measurements. From these observations, a more accurate map of Korea was produced based on the combined work of Jesuit cartographers and Chinese surveyors, no doubt using Korean maps as guides.

Hostetler (2001) argues that cartography was an important component of early modern empires and that, in their mapping exercises, the Qing rulers followed much the same path as the French and Russians. In mapping their expanding empire they employed the same map language as their competitors. There was a new territorial need; from 1660 to 1770 the Qing doubled the amount of land under direct control. All the expanding imperial powers of the day used international experts. The Qing mapping was less a unique Sino-Jesuit interaction than simply a case of international experts being employed by just one of many expanding powers eager to use cutting-edge technology to visualize, represent, and embody their power. Hostetler makes the case that the adoption of early modern mapping techniques in France, Qing China, and Muscovy was part of a global movement and that Chinese mapping was also part of this universal cartographic tradition. The imperial survey employed both Jesuit mapmakers and local experts.

Beginning about 1717, a large number of maps of the survey were published in a variety of forms. Hostetler (2009) considers four variants: a large scroll dating from

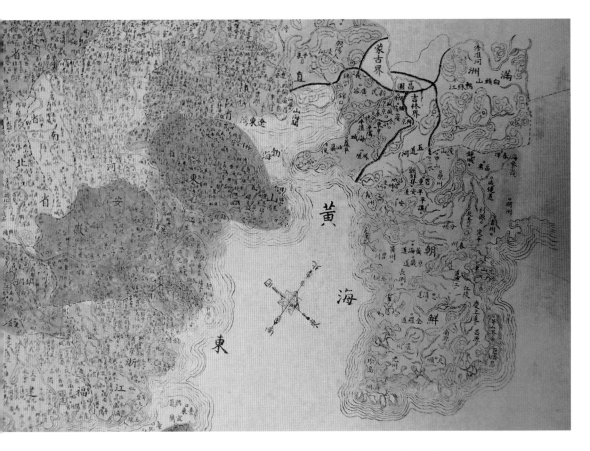

4.3 CHINA, 1885–94. (Courtesy of the Library of Congress, Washington, DC: G7820 1894. D3 Vault.)

1719, a 1721 atlas version of the survey, a French atlas published in 1737, and an edition of a Chinese-language atlas published in 1721. She considers each of them to be hybrid maps reflecting both indigenous cartographies and early modern practices.

From the late eighteenth century to the nineteenth, most Chinese mapmakers ignored these earlier influences. The voyages of Zheng He were forgotten, and the Jesuits eventually were banished from the country. China slid into a more parochial stance, trying to ignore the world outside its immediate sphere, and new world maps focused on China's centrality and preeminence. Korea continued to be represented, however, albeit on the outer edges of the empire. Figure 4.3 is part of a map of China made between 1885 and 1894, a woodblock print with relief shown by hachures. It shows the provinces of China as well as Taiwan, Korea, and part of Manchuria. It also has, as an insert, a double hemisphere map of the world. The hachures and the double hemisphere projection suggest a more modern cartography, but showing Korea as part of a greater China is consistent with a Chinese mapping tradition that extends back over seven centuries. The map thus represents the production of global space within more parochial Chinese imperial concerns.

MANCHURIA

Like all states, Joseon Korea was sensitive to border issues. The northern border with Manchuria was highly vulnerable territory and hence a significant object of cartographic representation. Maps of this area were an integral part of official Joseon mapping down through the years. Soon after the regime was established, it undertook extensive mapping of the northern border regions. The country's danger from its northern neighbors was demonstrated by the Manchu invasions of 1627 and 1636, a precursor to the fall of the Ming dynasty and the establishment of the Qing in 1644. Under the Joseon monarch King Hyojong (r. 1649–59) in particular, military bases were established and strengthened along the northern border. A ten-screen panel map, the *Yogyedo*, 139 centimeters by 635 centimeters, was completed in 1706 by Yi Imyong (1658–1722), who had traveled to Beijing on a diplomatic mission a year earlier and had obtained Chinese maps and books. He combined these with Korean and Manchu maps of the border regions to produce a large painted map that depicts a vast area from Beijing to the Pacific.

The northern border was mapped throughout the later Joseon period, even when the Qing had firmly established control and relations with China were relatively peaceful. Kang (2008) suggests that frontier maps made after the erection of a boundary marker on Baekdu Mountain in 1712 were more accurate than those made before this date. Figure 4.4 is a map of this border region made by an unknown Korean cartographer sometime between 1733 and 1858. It is very detailed, showing the military districts in Manchuria and noting the distances between the places depicted and Seoul. The subject matter and quality indicate that it is an official government map made to provide the central authorities with an accurate cartographic representation of this important border region. Korean towns and villages are marked as red circles along the river border. The map also has a city plan of the Shenyang, the political headquarters of the Manchu before they founded the Qing dynasty. Maps of this border region were necessary to let the leaders keep a steady cartographic eye on a close and powerful neighbor. The writing and mapping of the northern border became "a primary site for examining and even challenging the limits of Korean national space" (Schmid 2000, 219). Looking toward the north, especially in the early twentieth century, gave many Korean writers a dis-

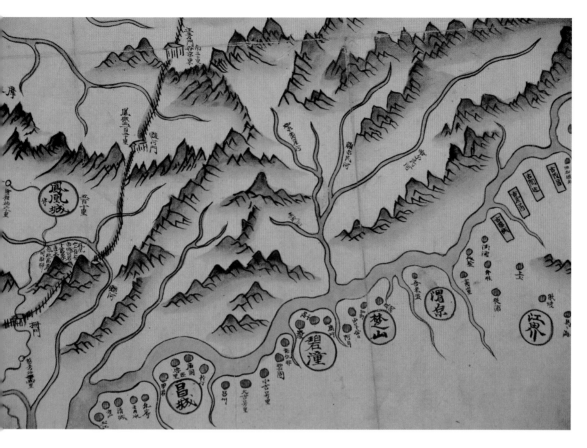

4.4 MANCHURIA, 1733–1858. (Courtesy of the Library of Congress, Washington, DC: G7822 M2 1747.S5.)

tinct sense of the containment of territorial expansion and the loss of an extended national space (Schmid 2002).

JAPAN

Maps played an important role in traditional Japanese society both as decorative pieces and as practical tools in a centralized hierarchical system of control and surveillance. In the eighth century the central government of Japan ordered maps to be made of the various provinces. These maps, drawn in a distinctive style, are known as the *gyoki* maps after a Buddhist priest named Gyoki (668–749). *Gyoki* maps generally show provinces, roads, and cities. They dominated Japanese cartography from the eighth to the nineteenth century. When the central authorities commissioned a survey of provinces in the seventeenth century, another series of *gyoki* maps were produced. Japanese society was deeply dependent on the rice crop, and accurate grid maps of rice paddies, also dating from the eighth century, were essential records of a very important resource, minimizing disputes over landownership. Japanese cartography tends to be more geometric and reveals very limited influence of the Korean shapes and forces form.

The 1402 *Gangnido* contains a representation of Japan that draws on a *gyoki* map of Japan obtained by the Joseon diplomat Pak Tonji, who visited Japan between 1398 and 1402. The direct use of the *gyoki* map highlights the cartographic relations between the two countries. There was a long interchange between Korea and Japan as ideas and techniques bridged the narrow seas that divided the two countries. Shared traditions of Buddhism came to Japan by way of Korea in the sixth century, and both Japan and Korea had a Buddhist legacy of world maps. But they never had the same depth of cultural affinity that linked Korea with China. Japan was outside the tight cultural and political nexus that bound Korea to China—more overtly independent.

Yet the proximity of Japan and Korea did lead to maps' being made of the neighboring countries. A 1482 list of Korean maps includes a map of Japan and Ryukyu (Okinawa) made in King Sejo's reign, between 1455 and 1468. Later, King Seongjong ordered a book to describe the neighboring lands. The 1471 *Haedong jegukgi* (Chronicle of the Countries in East Asia) contains several maps of Japan. Bae (2008) suggests that the primary source was maps made by Japanese who had traveled to Ryukyu and surrounding islands.

Maps of Japan regularly appeared in Joseon atlases from the seventeenth century, and Edo maps were introduced in the eighteenth. Intellectuals such as Jeong Yak-yong were especially interested in Japanese geography and had copies of maps of Japan. The closeness of the two countries also led to specific forms of mapping. The Japanese invasions of Korea in the last decade of the sixteenth century meant the Joseon dynasty never quite trusted its Japanese neighbors. Although there was relative stability in Joseon-Tokugawa relations from 1600 to 1868, the Koreans kept a careful eye on Japan. Defense maps were made of the coastlines of the southern provinces of Cholla and Kyongsang, not only because of the distant fear of invasion but also owing to the immediate threat of Japanese pirates. An early nineteenth-century map, *Map of the Natural Defenses of the Southern Coasts of Kyongsang and Cholla Provinces*, shows military bases, towns, and harbors along the entire southern tier coastline. The map supplied detailed military and naval intelligence in case of piracy or Japanese attack.

Japan also featured in the Joseon worldview. Along with China, a map of Japan made a regular appearance in the atlases of the late Joseon. Figure 4.5 is an atypi-

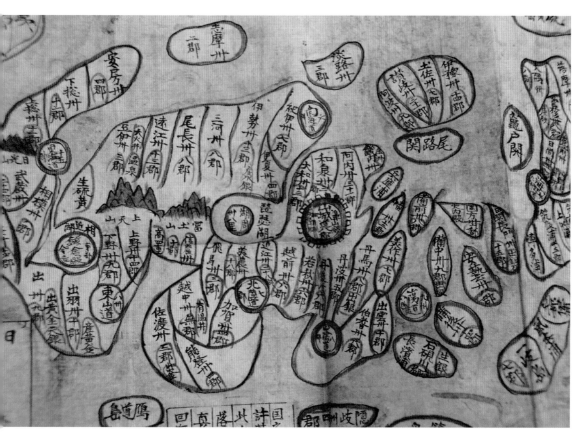

4.5 MAP OF JAPAN, nineteenth century. (Courtesy of the Library of Congress, Washington, DC: G2330 Y667 18— Vault.)

cal map of Japan in a small hand atlas; it is placed alongside a world map, a map of China, and maps of Korean provinces. This map is very distinctive—more in the Japanese geometric style—and suggests a more direct Japanese source more than a Korean map of Japan.

Whereas Japan figured in Joseon maps, maps of Korea rarely figured in Japanese cartography. The cartography of the early Tokugawa period was mainly devoted to Japan. Berry (2007) argues that this cartography helped create a sense of nationhood, national identity, and national belonging. The singularity of a national cartographic representation helped in achieving a singular national identity. This narrow cartographic provincialism has continued down to the present (Sato 1996). The central authorities undertook five complete sets of provincial maps in 1605, c. 1633, 1644, 1697, and 1835. The maps displayed the power of a regime seeking to stabilize and extend its rule. Yonemoto (2003) makes the case that by the late seventeenth century and especially in the eighteenth, printers and artisan mapmakers inherited spatial concepts and map visions and transformed them for the commercial market. The new print culture led to vernacularizing national space for new audiences. These new maps gave a wider angle on the usual national depiction. Unlike most official maps, the commercial maps showed other countries. The *Outline Map of Our Empire* by Ishikawa Ryusen, printed in 1687, consigns Korea to the edge of the map. Although the central focus is on national representation, even this marginal depiction gives a sense of Korea's proximity.

Japanese cartography was transformed by changing external relations. First there was the contact with China. Later, Europeans moved into the region. The early European traders were called *nanbanjin*, meaning "southern barbarians." *Nanban* maps show the world on projections learned from the Europeans, such as the Mercator projection. In a fusion of East and West, the *nanban* maps are an example of intercultural cartography, as are Japanese marine charts from the seventeenth century onward, which were influenced by Portuguese maritime cartography. Even the word for map was modified in the cultural exchange. Potter (2001) reviews the term's semantic history to find that the Japanese language had no distinctive word for map. The long-used *zu* referred to diagrams and referenced the strong pictorial element in early Japanese cartography. Maps with pictorial information tended to be more highly prized far into the eighteenth century. The use of *chizu*, a neologism literally meaning "land diagram" and closer to the Western notion of map, emerges only in the latter half of the eighteenth century in the wake of contact with European maps and mapmaking techniques. A similar linguistic transformation occurs with the Chinese word for map.

Then, in the later Tokugawa period, the country also felt the press of expanding empires. Beginning in the 1780s the Russian empire extended its reach into the North Pacific, and this incursion, in association with the beginnings of Japanese imperialism, prompted more accurate mapping of the wider region. Walker (2007) notes the Japanese exploration and mapping of the Sakhalin Island in the very early nineteenth century. In 1808–9 the Japanese cartographer Mamiya Rinzo (1775–1844) mapped the island, demarcating the borders between Russia, China, and Japan. Walker argues that the mapmaker anticipated empire by figuratively emptying the land of inhabitants and capturing the territory in a grid of empire and nation.

There was a strain in the ideology of the Tokugawa regime. Trade monopolies with the outside world created huge profits for the shogunate, but the authorities tried to limit the influence of the foreign traders. After 1635 ships were able to enter Japanese waters only from China, Korea, and the Netherlands, according to a strictly controlled system. The authorities wanted to benefit from foreign trade yet maintain control by restricting free entry of foreigners and denying foreign travel to the Japanese people. The shogunate wanted to remain self-contained and to be left alone. Although the policy of seclusion and parochial self-containment was dominant, there were those who voiced the need for a more activist presence in the region. Figure 4.6 is a 1785 Japanese map of Korea, in a distinctly Japanese style. This pen-and-ink watercolor was probably a copy of a manuscript map published by Hayashi Shihei (1738–93), a scholar of military strategy who promoted Japanese military power. With hindsight, we now see him as a forerunner of a more expansionist Japan. Eager to build effective military competence and a strong maritime presence, in 1796 he published *An Illustrated Description of Three Countries,* which deals with Korea, Okinawa, and Japan. One of his purposes was to demonstrate Japan's proximity to its neighbors. His writings highlight the external threats and the need to build more effective military and naval power. The map of Korea shown in figure 4.6 is an important document in the beginnings of a more expansionist Japan.

EUROPE: A NEW NEIGHBOR

With the globalization of the early modern period, Korea and Europe moved closer. There were growing connections, shared practices, cultural exchanges, and scientific translations—slow at first, but building through the years as a shared global space was created and represented.

Figure 4.7 is a map of the world made by the famous Venetian cosmographer Vincenzo Coronelli (1650–1718), a Franciscan scholar who made globes and maps for the ruling elites as well as for a wider market. As the royal cartographer to Louis XIV and cosmographer to the Republic of Venice, he was at the center of a worldwide network of shared geographical knowledge. In 1693 he published *Libro dei globi,* a set of maps that form the basis for one of his terrestrial globes. The book was destined for an international market and thus represents an up-to-date and widely diffused European-based picture of the world at the end of the seventeenth century. Figure 4.8 shows in closer detail the depiction of Korea in the 1701 edition of the book. Notice that Korea is now represented as a peninsula, and the coastal confusion of earlier European maps is resolved. Coronelli drew on wider and deeper European geographical knowledge of the area to represent the country more accurately, at least in basic outline. But also notice that there is little annotation for Korea. The interior of the country remains largely blank. Compare this with the details of China, where the Great Wall is clearly visible, though hardly accurate. The coastline around Japan is also marked with many names. China and

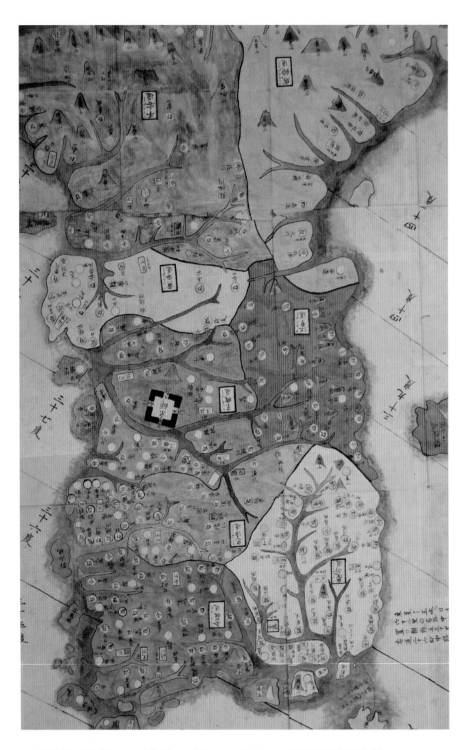

4.6 KOREA, 1785. (Courtesy of the Library of Congress, Washington, DC: 7900 1785 .H3 Vault.)

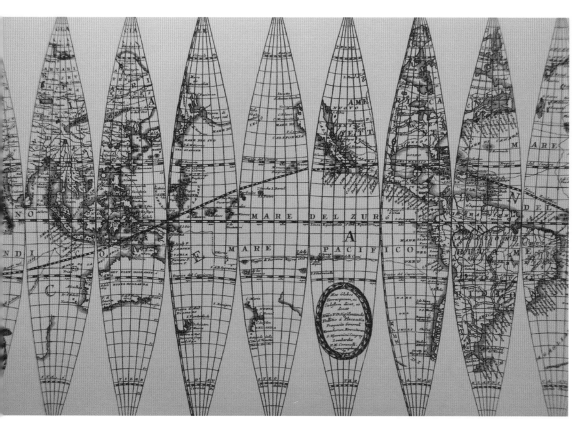

4.7 WORLD MAP, 1701. (Vincenzo Coronelli, *Libro dei globi* [Venice: Gli Argonauti, 1701]. Courtesy of the Library of Congress, Washington, DC.)

Japan are now relatively well known to European cartographers, and there is a sharing of cartographic knowledge between East and West. We are now in the era of cartographic encounters between China and Japan, on the one hand, and Europeans on the other. Korea, however, is still on the margins of this connection. The outline of country is better known, but the details remain sketchy.

Cartographic information on Korea came to Europe through two main sources. The first was the Chinese and Japanese, who had some knowledge of the kingdom. Early Chinese and Japanese maps portrayed the peninsular kingdom more adequately than the often hypothetical accounts in early European maps. With the permanent entry of the Europeans into the Far East, a complex cartographic interaction began. It took a variety of forms, including Sino-Jesuit maps made by European Jesuits in the service of the Chinese government; Chinese and Japanese maps

4.8 DETAIL FROM CORONELLI'S MAP. (Vincenzo Coronelli, *Libro dei globi* [Venice: Gli Argonauti, 1701]. Courtesy of the Library of Congress, Washington, DC.)

that drew from European sources and employed new cartographic techniques; and European maps that used Chinese, Japanese, and Korean sources. Let us consider examples of all three.

The first is Ferdinand Verbiest's 1674 map of the world, produced for a scholarly audience (Walravens 1991). We have already discussed Matteo Ricci's map. Verbiest was born at Pittem in Flanders and became a Jesuit priest at age eighteen. He traveled to Macao in 1659 and later moved to Peking, where he immersed himself in the Chinese culture, taking the Chinese name Nan Huairen (Nan Huai-jen). He was an adviser to the emperor and made contributions to astronomy and cartography. The Verbiest map of the world, made at the emperor's request, is a double hemisphere woodblock originally printed on silk, titled *Kunyu quantu* (Map of the Whole World). A detail from this map in figure 4.9 shows a peninsular Korea. Verbiest's map is a good example of collaborative transcultural mappings: Sino-Jesuit maps made by European Jesuits at the Chinese court that synthesized European and Chinese cartographic traditions to more accurately represent Korea to the educated public.

The second example is a Japanese screen map from the seventeenth century now held at Idemitsu Museum of Arts in Tokyo. Titled *World Map and Representations of Forty Nationalities,* this large Japanese-Dutch world map uses the Mercator projection (a European projection) and draws on Dutch knowledge. The Dutch arrived in Japan in 1600, but they were not the first Europeans. The Portuguese arrived in 1543 and brought not only firearms, quickly appreciated by the authorities, but also Catholic missionaries. The missionaries were initially welcomed, but by the late sixteenth century the Japanese leaders considered them a threat to the social order and banned them in 1587, when the total Christian population was estimated to be about 300,000. The Dutch, more concerned with trade than converts, were restricted to the tiny island of Deshima in the large harbor of Nagasaki, and by 1639 they were the only Europeans allowed to trade in Japan. Through this small aperture, regular Dutch sailings brought in silk from China and foods from Europe and sailed off with Japanese ceramics. Maps were also shipped in to meet a growing Japanese demand and were prized as much for their art as for their geography. Japanese artists used these Dutch maps to create their own versions, often painted on screens. Figure 4.10 is a fragment of the large map that highlights Korea. The English writing is very recent and not part of the original map.

The third example is a French map from the 1708–17 Sino-Jesuit mapping of the Qing empire. Some of the maps were sent to France and published in Jean-Baptiste du Halde's 1735 *Description de l'empire de la Chine at de la Tartarie chinoise,* one of the most comprehensive accounts of China for a Western audience. The book contains forty-two maps based on the 1708–17 surveys. Scales are given in

4.9 DETAIL FROM VERBIEST'S MAP OF THE WORLD, 1674. (Ferdinand Verbiest, *Kunyu quantu*, 1674.)

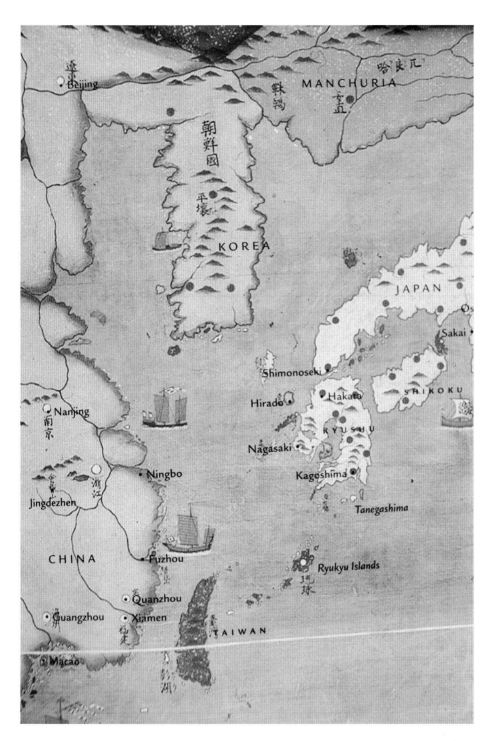

4.10 JAPANESE WORLD MAP, seventeenth century. (Idemitsu Museum of Arts, Tokyo.)

4.11 DETAIL FROM MAP OF KOREA. (Jean-Baptiste d'Anville, *Nouvel atlas de la Chine, de la Tartarie chinoise, et du Thibet* [La Haye: Scheurleer, 1737]. Courtesy of the Library of Congress, Washington, DC.)

Chinese and French measures. France's most accomplished cartographer, Jean-Baptiste d'Anville, drew the maps. One is titled *Royaume de Corée* (Kingdom of Korea). The same map was also published two years later in d'Anville's *Nouvel atlas de la Chine, de la Tartarie chinoise, et du Thibet,* which included all the maps in the Du Halde volume. The d'Anville atlas is one of the most detailed representations of China and its empire for a European audience; there are fifteen provincial maps of China, twelve of Tartar China, and even nine separate maps for Tibet. Yet Korea only has one map. Figure 4.11 is a section of that map that depicts part of the southern coastline. This was the most accurate map of the Korean kingdom to date, not superseded for almost two hundred years. It is a pivotal work in the cartographic representation of the country. The d'Anville map is a signal achievement in the mapping of Korea that combines the work of Jesuit cartographers and Chinese surveyors, no doubt using Korean maps as guides, published in Paris and disseminated widely through Europe in many subsequent editions and copies. For the first time, Western audiences had a fairly accurate map of Korea.

Korea was relatively closed to Europeans throughout much of the entire Joseon period, from 1392 to the very late nineteenth century. However, it was still being mapped by European powers as part of the production of global space. Mapping the world in all its senses, including the cartographic depiction of the globe as well as the description and classification of foreign parts of the world, was an integral and explicit part of the Enlightenment. In England the Royal Society was established in 1660, and the French equivalent, the Académie des Sciences, was created in 1666. There was a close connection between the extension of geographical knowledge and the Enlightenment project (Livingstone and Withers 2000). Geographical issues had a central place in the Enlightenment preoccupation with reason, rationality, and science. The world was an exhibition to be mapped and described. "The Great Map of Mankind is unroll'd," wrote Edmund Burke in 1791. This "unrolling" was most evident in the range of ostensibly scientific surveys sent out in the eighteenth century to map the vast Pacific. The grand explorations of Britain's Captain Cook, the Pacific voyages of France's La Pérouse and d'Entrecasteaux and Spain's Malaspina expeditions were acts of both grand geopolitical positioning and scientific curiosity. The expeditions contained illustrators, artists, scientists, and mapmakers who sought to describe, classify, and map as they made their way around the world.

The desire for increased scientific knowledge was deeply bound up with the search for commercial opportunities and the promotion of national economic interests. Letters written in 1782 between naval personnel and well-connected politicians about James Cook's last voyage, for example, reveal this concern with commercial possibilities: "Extensive trade that may be carried out to the N. E. of China, the Kingdom of Korea, to the Lius Kiu's and several islands in these seas. Those countries laying in Northern latitudes and exposed to very severe Winters will naturally demand several articles of our Woolen manufacture, and introduce a very beneficial balance of trade" (cited in Fry 1973, 189).

In 1769 Alexander Dalrymple, a geographer, a member of the Royal Society, a writer for the East India Company, and later the first hydrographer of the British Admiralty, published a book titled *A Plan for Extending the Commerce of This Kingdom and of the East-India-Company*. He notes that Korea "is extremely populous, it is in extent equal to France, no Europeans have ever carried on a direct trade" (Dalrymple 1769, 92–94). Dalrymple, as writer, promoter of economic expansion, chartmaker, and commercial propagandist for free trade, embodies the links between science and commerce, between exploration and the geographical description of the Enlightenment project, and between scientific knowledge and commercial opportunity.

The European mapping expeditions to the Pacific were part of national projections of power as well as expanding national commercial opportunities. We can consider three examples from different countries—France, Russia, and Britain—that involve Korea. The first is the exploration of La Pérouse or, to give him his full title, Jean-François de Galaup, comte de La Pérouse, a French naval officer who in 1785 was appointed by Louis XVI to lead an expedition around the world. He was charged with undertaking geographic and scientific research, keeping an eye open for commercial opportunities, and flying the French flag around the globe. The expedition was part of the national rivalry with Britain and in particular was an attempt to emulate Britain's Captain James Cook, whose successful round-the-world voyages, the first from 1768 to 1771 and the second from 1772 to 1775, did much to rewrite science, reposition geopolitical alignments, and promote British interests. The published accounts of Cook's journeys captured the European imagination. The La Pérouse expedition was the French response to Cook, since they had lost their colonies in North America and were eager to expand into the North and South Pacific. La Pérouse left Brest on August 1, 1785, with two ships, *La Boussole* and *L'Astrolabe,* and sailed around South America and into the Pacific. He reached Macao in January 1787, then sailed north, passing the island of Jeju between Korea and Japan, close to the Korean coast, and the island of Ulleungdo (La Pérouse called it Dagelet Island, after one his officers). The strait between Sakhalin and Hokkaido is still known as La Pérouse Strait. He sailed south to Botany Bay, arriving just after the British, and stayed six weeks in the fledgling colony. He left Australia, sailing north, and was never heard from again. But his journals and some of his maps and charts survived, since he transferred them at various points in his journey. Figure 4.12 shows a detailed map of Ulleungdo made by La Pérouse and published posthumously (Dunmore 1994). When the ships came across the island, the captain noted that it did not appear on any of his charts. So on May 27, 1787, he tried to approach but was pushed back by prevailing winds. The next day the wind dropped, and La Pérouse was able to sail around the island. Unable to find water deep enough to anchor, he lowered a small boat that enabled the French to take more detailed measurements. In his journal La Pérouse noted the exact reference point as "the Northeast point of the island is 37 degrees 25′ North latitude and the 129 degrees 2′ East longitude." Korea was being placed on the universal grid. Although he was unable to land or even converse with the locals—Koreans who came to the island every summer to build boats that they took back to the mainland for sale—La Pérouse, because of the accuracy of scientific instrumentation, was able to plot the island's

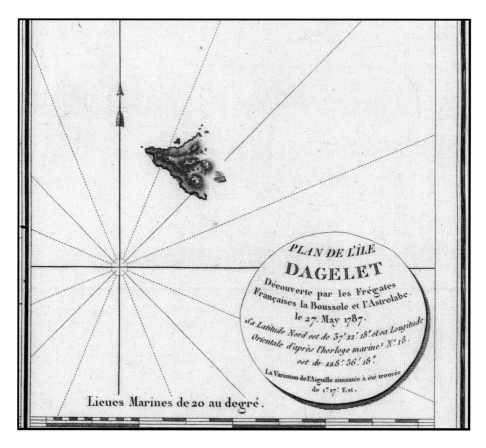

4.12 DAGELET ISLAND, 1788. (Jean-François de Galaup, comte de La Pérouse, *Voyage de La Pérouse autour du monde* [Paris: Imprimerie de la République, 1797]. Courtesy of the Library of Congress, Washington, DC.)

position. Korea, despite its insularity, was gridded and plotted as part of a universal mapping endeavor, being incorporated into the production of global space and becoming part of the modern world.

Korea also appears on a number of Russian maps. The Russian empire was expanding its dominion eastward. From 1550 to 1700 the small state of Muscovy became a continental power, extending its reach to the Pacific. The lure of the fur trade and the need for geopolitical positioning pulled Russia farther east. Peter the Great hired the Danish sailor Vitus Bering to explore the Far East. He led two expeditions, one to Kamchatka in 1728 and one to North America in 1741. A map of the first voyage was published in the d'Anville *Atlas* of 1737.

As an expanding imperial power, the Russians mapped the empire under their control as well as adjacent territories (Kivelson 2006). In 1696 the czar commissioned a map of Siberia that also shows the "lands of enemies." Korea appears for the first time between 1699 and 1715 on Russian maps in atlases produced by Semen Remezov, who composed many maps of Siberia (Bagrow 1958). The Russian maps depended on explorations in the region and on contact with the Chinese. Tolmacheva (2000) identifies distinct periods in Russo-Chinese diplomatic relations that resulted in cartographic advances, including the intelligence gathering leading up to the signing of border treaties in 1689 and 1727 and to the exploration campaign of 1848–63.

Russian influence in what was for them the Far East waxed and waned. After an initial period of expansion that led to the border treaty of 1689, they largely withdrew from the area, only to begin a more expansionist policy in the 1840s. The Amur region was annexed in treaties signed with China in 1858 and 1860. This nineteenth-century expansion is examined by Mark Bassin (1999), who argues that the very remoteness of the area made it susceptible to numerous projections. Notable was the depiction of Amur as the Russian equivalent of the New World and thus a vessel for Russian nationalist sentiment as well as a vehicle for the discursive universal mission of conveying civilization. Even the radical intellectual Alexander Herzen wrote of the Russian expansion and annexation of Manchuria as "one of civilization's most important steps forward."

Through the intellectual circuits that connected St. Petersburg, Amsterdam, Paris, and London, the maps from the edge of the Russian empire were assembled and published throughout Europe. *Carte nouvelle de tout l'empire de la grande Russie* (New Map of the Great Russian Empire), published in Amsterdam in 1724, contained depictions of Korea. The Imperial Academy of St. Petersburg published a series of maps in 1775 that were widely available in Europe. A map of Korea appears in the four-volume atlas of 214 maps printed in 1779 by Antonio Zatta, titled *Atlante novissimo*.

The Russian navy was also involved in exploring the North Pacific. In the early years of the nineteenth century, from 1803 to 1806, Ivan Fedorovich Kruzenshtern, commanding the *Nadezhda*, one of two ships undertaking a circumnavigation in the style of Cook and La Pérouse, sailed in the western seas of the North Pacific. Later sailing voyages by Otto von Kotzebue in 1816–18 and again in 1823–26 and by Fedor Lütke from 1826 to 1829 all added to geographical understanding of the wider region. Figure 4.13 is a map of Korea that was included in the *Atlas iuzhnogo moria* (Atlas of the Southern Sea) by Captain Kruzenshtern that was published in St. Petersburg in 1826. The map has little information of mainland Korea. No roads are shown; no towns are listed. It is a map made from the sea: the offshore islands and coastline are the main focus.

As part of the wider regional cartographic representation of the region, the

4.13 RUSSIAN MAP OF KOREA, 1826. (Ivan Fedorovich Kruzenshtern, *Atlas iuzhnogo moria* [St. Petersburgh: Sankt, 1826]. Courtesy of the Library of Congress, Washington, DC.)

Russians mapped Korea as they explored the limits of their empire and beyond. Even as late as 1900 the Russian Geographical Society published the report of an exploring expedition through Korea led by a member of the society, M. Schmidt. The primary object was studying flora and fauna, but as reported in the *New York Times* of November 11, 1900, "He took advantage of the opportunity, however, to make topographical observations."

As the preeminent naval power for almost a hundred years, from 1800 to 1900, the British navy played an important role in producing global space by plotting coasts, islands, seas, and oceans throughout the world. William Broughton was part of this global endeavor. Broughton was a captain in the British navy who worked on surveys in the Northwest Pacific. In 1796, as commander of the four-hundred-ton sloop HMS *Providence*, he undertook a two-year mission that surveyed the coast of Asia from Sakhalin Island to the Nanking River. He wrote up his journey in a book published in 1804, *A Voyage of Discovery to the North Pacific Ocean*. Broughton's journal is a dry account, full of positional readings as places were incorporated into the universal grid of latitude and longitude. He lives on through various place-names. Many subsequent Western maps name the sea between Korea and Tsu-shima "Broughton Straits," and a variety of "Broughton Bays" still appear on maps of the southeast coast, the northeast coast, and the west coast of Korea.

By October 1797 Broughton sailed past the Japanese island of Tzimu (Tsushima) and on to the southeast coast of Korea. "As we drew in with the land we observed several villages scattered along the shore" (Broughton 1804, 328). He went ashore and visited a local village, probably Chosan (Pusan), making scientific readings of altitude, latitude, and longitude; his readings gave the location as 35 degrees, 2' north, 129 degrees, 7' east. "After taking some altitudes for the watch, and observ-ing the distances for the longitude, we took a walk, attended by a numerous party of villagers. . . . Many villages were scattered round the harbour and in the N.W. part we observed a large tomb encircled with stone walls and battlements upon them" (Broughton 1804, 330–31). Later that same day a group of "superior peoples" visited the British party. "They were dressed in large loose gowns, with very wide brimmed high crowned black hats each with fan. Their inquiries seemed to tend to a knowledge of what brought us to their country; but I fear our replies gave them very little satisfaction, as we would so little comprehend each other" (Broughton 1804, 332). The British returned to their ship, but the next day a boat brought even more "superior people" who, despite the language difficulties, made it plain they wanted the British to leave. They even sent out water and wood to hasten their departure.

We can imagine the way news of the foreigners moved through the political system, involving officials ever higher up the hierarchy. The officials, Broughton noted—their status finely graded by the quality of their attire—wanted little con-tact and were keen to make the British move on. Incomprehension and indifference marked the human encounter. But also figuring in the contact were Broughton's careful measurements of altitude, latitude, and longitude. It was mathematical rather than cultural mapping. Korea was measured and incorporated as a space rather than being understood as a society. Although the Koreans were unwilling to interact, scientific measurement was taking place nonetheless, incorporating their coastline into modern maps. It is an apt metaphor for Korea's entry into the modern world: desiring separateness, the inhabitants nonetheless were unable to avoid incorporation into the unfurling map of modernity.

5

CARTOGRAPHIES OF THE LATE JOSEON

The long-lasting Joseon dynasty is divided in a number of ways by historians. Traditionally there is a threefold distinction: the early Joseon of the first hundred years was marked by competent rulers committed to centralizing political control. Then came a middle Joseon comprising almost two centuries of invasion and war with China and Japan, followed by the late Joseon, characterized by a weakening of Confucianism and the traditional hierarchical society and a long line of ineffectual rulers, apart from Yeongjo and his son. In this chapter I will adopt a simple division between early and late Joseon. This distinction is not precise. There are enough similarities to see the two phases as a single unit, one of the longest continuous regimes in the world. The state functioned as a coherent whole during the entire period. However, several things do distinguish the early Joseon era from the later one, allowing us to make the distinction.

First, the need to legitimate the regime, especially important at its beginnings, became less significant as time passed. Before the Joseon dominance, Buddhism and Confucianism played complementary roles, with the former serving as the state religion while the latter shaped the political sphere. Buddhism was undermined during the early Joseon, and temple lands were confiscated. The Joseon claimed legitimacy by reviving indigenous traditions, suppressing Buddhism as the state religion, strengthening links to China as the center of world civilization, and establishing a Neo-Confucian social and political order. As time passed and centuries of Joseon rule continued unabated, the emphasis shifted from creating legitimacy to maintaining it.

Second, in the later Joseon era there were unprecedented external political challenges. The sixteenth-century entry of European powers into the region dramatically changed relations between Korea and the outside world, because calculations no longer included only the immediate neighbors China, Japan, and Manchuria. The Dutch, the British, and later the Americans played a growing part in the geopolitics of the region as foreigners brought new ways of seeing, imagining, and representing the world that filtered into Korean cartographic practices. And then there were changed relations between the countries of the region. About 1600 there were tremendous upheavals, followed by a new stability that was to last until the latter half of the nineteenth century.

A break between the early and late Joseon occurred about 1600, when Korea became a battleground between Japan and China. After unsuccessfully seeking a Korean alliance against China, Japan invaded the country in 1592 with 160,000 troops. A second attack in 1597 by 140,000 troops wrought further damage. Troops of the Ming empire also fought against the Japanese in the Korean peninsula, further devastating the country. The Manchu, who went on to invade China and establish the Qing dynasty (1636–1911), invaded Korea in 1627 and again in 1636. After almost fifty years of invasion, war, and social turmoil, it is understandable that the Joseon established a policy of seclusion, restricting relations with both China and Japan and avoiding contact with the Europeans. The policy of seclusion was largely successful for almost two hundred years because of relative dynastic stability in the region. The Qing dynasty lasted from 1644 to 1911, and the Tokugawa shogunate lasted from 1600 to 1868. For over two hundred years then, from the mid-seventeenth century to the mid-nineteenth, there was relative stability in the governing regimes of China, Japan, and Korea.

Third, although the later Joseon, until its very last phase, is noticeable for the stability of its external relations, it was also marked by growing internal social tensions. Traditional Joseon society was hierarchical. Males over sixteen had to carry identification tags, called *hopae*, recording name, date of birth, and social status. This rigid system was founded on the assumption of stability and an unchanging social order. But population growth, economic development, and associated social change all created new forces that threatened to undermine the careful stability of this Neo-Confucian society. There was also the recurring problem of funding the state, since the landholdings of the aristocrats were exempt from taxation. State revenues largely rested on land taxes, which could impoverish the peasantry in weak harvest years.

The weight of settled tradition confirmed and reproduced the strict social hierarchy marked by a very sharp apex of royalty and aristocrats and a wide base of commoners. Confucian principles required that government administration be a meritocracy, but in Korea governing positions were restricted to upper-class males. The well-placed scholar-officials of the landed aristocracy were called the *yangban*. Their emphasis was on learning rather than on military pursuits, so a military class did not emerge in Korea to the same extent as in neighboring Japan. The traditional *yangban* also tended to eschew commerce as beneath their dignity. This bureaucracy was very conservative, unable to change quickly or react to rapid and unprecedented political changes, and was more concerned with self-enrichment and maintaining social position than with advancing the public good. Factionalism between aristocratic families was also a source of tension, as was the recurring problem of the peasantry, whose living conditions could deteriorate sharply in years of meager harvests. There were major peasant uprisings and numerous local rebellions across the entire period, all fueled by mass starvation. Almost a million people died of famine and associated diseases in one eighteen-year period in the nineteenth century.

Against the general backdrop of an oppressed rural population occasionally on the verge of starvation, a self-serving bureaucracy, and a fractious ruling elite, there was the growth of new social classes and attitudes. Economic growth and trade flourished under the political stability of the country and the region. Some farmers became wealthy, merchants were enriched, and a new commercial spirit contested the Neo-Confucian emphasis on learning rather than commerce. There was also the beginning of upward social mobility as the commodification of agricultural markets and economic growth enabled richer peasants and canny merchants to rise from the bulk of commoners. By the eighteenth century, such things as the production of more genre paintings and everyday porcelain are evidence of a rising affluence among selected merchants and farmers. A more critical spirit displaced unthinking deference to the *yangban*. Growing literacy widened the political discourse.

Fourth, there was an epistemological rupture caused by the entry of new knowledge. Just as medieval and early modern geographical understanding in Europe was transformed by a maps that included the New World, so Joseon Korea was affected by new maps that replaced China as the center of the world. From the early seventeenth century, Korean emissaries to China brought back Sino-Jesuit world maps undermining the old certainties with new discoveries about the shape and size and configuration of the world. The maps created a marked epistemological shift by fundamentally questioning traditional learning and by undermining long-established tenets. The new knowledge was contested as well as embraced.

Fifth, an important theme in the history of ideas in the late Joseon was the idea of *silhak*, often translated as practical learning. The emphasis on purely meta-physical deliberations, so strong among the traditional *yangban*, was challenged from the late seventeenth century onward by scholars eager to use knowledge and learning to solve pressing social problems. In the seventeenth and eighteenth centuries, concern focused on the problems of rural areas—improving agricultural productivity and creating a more equitable system of land taxation. Later, proponents of industrialization and the introduction of foreign technology sought to modernize the country's economy. As social tensions increased, this emerging practical school of learning sought solutions for more economic, social, and technological issues. The practical learning school sought to raise public awareness and develop social policies. Its ideas were promoted especially during periods of crisis, when both breakdowns in the economy and questions of legitimacy provoked wide-scale criticism of the status quo. *Silhak* was a growing counterweight to the aloof *yangban*'s disconnect from practical considerations and their failure to address urgent policy issues dealing with the lives of the general public. The *silhak* movement never achieved dominance, but it did influence the tenor of debates. Its proponents argued for a more fact-based empirical approach and for cultivating a national spirit by encouraging the Korean language and recognizing Korea's unique history and identity. As we will see later in this chapter, the movement had an effect on geographical knowledge and national mapmaking.

This later era saw various attempts to improve society. Under King Yeongjo (r. 1724–76), for example, a new policy of taxation was introduced and agricultural productivity was encouraged. Various reformist schools of thought emerged to argue for a more efficient government, a more equitable society, and a more productive economy. The every existence of these schools was a symptom of the economic and political crises. With the benefit of hindsight, we see the late Joseon as the beginning of the end, leading to the collapse of the regime. We know from our historical vantage point that the regime's days were numbered and invariably see signs of decay and decline that were probably not so obvious to those experiencing them. It is clear, however, that a new commercial spirit and a more social critical attitude marked the later Joseon as economic change and demographic growth created new social forces that the traditional authorities sought to contain and control. Beneath the calm surface of uninterrupted Joseon rule was the increasing turbulence of social change and political conflict.

Commentators often describe the Joseon period from 1400 to about 1600 as one of peace, vigor, and cultural renaissance compared with the later period of growing internal conflict. The vigor of creating a Neo-Confucian society, a new alphabet, and new forms of imagining, seeing, and writing the country is often personified in the rule of King Sejong (r. 1418–50) who created groups of scholars to advise on the foundation of the state and to promote the advancement of science and technology. After about 1600, according to Haboush (2009), the monarchy and ruling elites never regained the self-confidence of the first two centuries.

MAKING MAPS

The cartography of the late Joseon both continues traditional models of mapping and creates new forms. It resists the production of global space but also is part of it. The Neo-Confucian social order stressed reverence for tradition rather than encouraging innovation. It is no surprise then that map forms created and developed in the early Joseon continued to be produced. Figure 5.1, for example, is a map of Korea, part of a manuscript atlas drawn about 1721. The cartographer is named Won, and all we know about him is that he had passed the civil service examination but had not yet been appointed to an official position when he compiled the atlas that contains this map. His atlas draws from the maps of China drawn by Zhu Siben (1273–1355) in the early fourteenth century. This map of Korea also reflects the style of a Chong Ch'ok map first developed in the fifteenth century. In other words, a Korean cartographer making an atlas in 1721 used Chinese maps originally made almost four hundred years earlier and mirrored the style of a national atlas developed almost two hundred years before. As the author writes in the preface,

> From my youth on I had the ambition to travel, but could not afford to wander over the 300 counties of Korea, much less the whole world. So carrying out an ancient practice, I drew a geographical atlas. And while gazing at it for long stretches at a time I feel as though I were carrying out my ambition. . . . Morning and evening while bending over my small study table, I meditate on it and play with it—and there in one vast panorama are the counties, the provinces and the four seas and endless stretches of thousands of miles. (Cited in McCune 1982, 21)

It is revealing that the "whole world" of this atlas consists of just China, Manchuria, and Korea. The preface also manifests the scholarly, almost meditative, element in late Joseon mapmaking; maps could be objects meant to provoke contemplation, even reverie, more than geographical tools to enable travel, commerce, and control. Maps functioned as mandalas as much as geographical information systems. The map of Korea shown in figure 5.1 highlights rivers, mountains, provincial boundaries, principal cities, and administrative centers. With its careful depiction of Baekdu Mountain on the northern reaches of the country and its highlight-

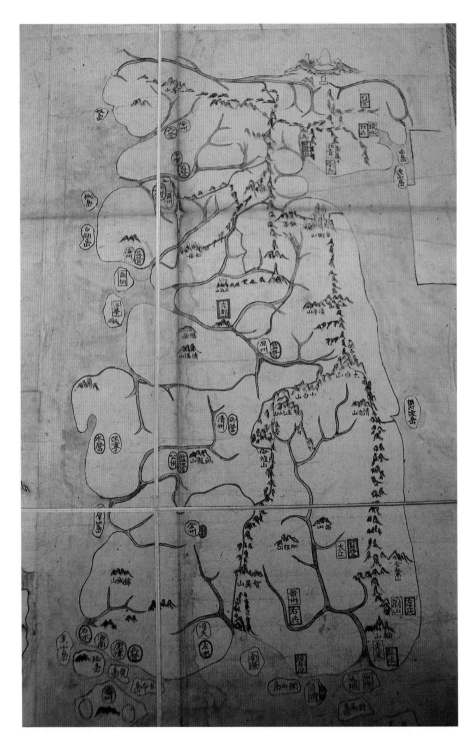

5.1 KOREA, 1721. (Courtesy of the Library of Congress, Washington, DC: G2305 D35 1782 Vault Shelf.)

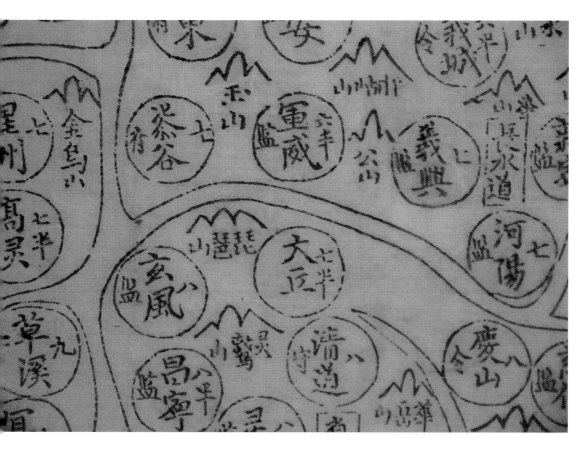

5.2 KYONGSANG, c. 1800. (Courtesy of the Library of Congress, Washington, DC: G2330 C53 18— Vault.)

ing of seats of government, the map links physical and political geographies into a coherent national whole, creating a one-to-one reading between enduring physical features and stable political structures. Looking at the map gives a sense of national coherence; representing sacred sites and centers of government together natural-izes the political and politicizes the physical.

Figure 5.2 is from a woodblock scroll atlas dated about 1800. It depicts a small part of a larger map of Kyongsang Province. Although produced sometime in the nineteenth century, it is in the style known as *sungnam*, first used in fifteenth-century China. The map is dominated by a typography that only hints at moun-tains and rivers. More like a written text or a scroll than a map, it represents the important role of calligraphy in Joseon cartography. Calligraphic characters are discrete units both representational and abstract and thus are easily adapted to a cartographic context, where they function as both visual and linguistic signs. Cal-

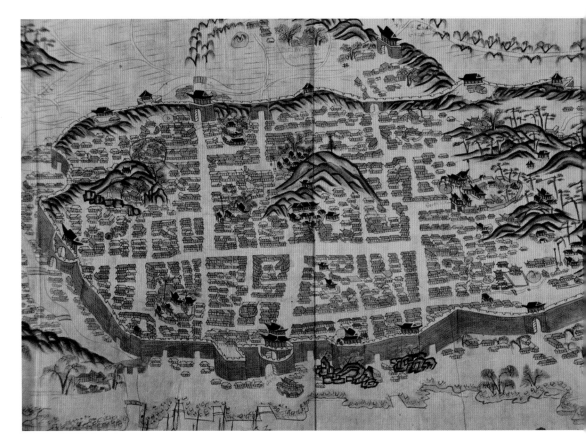

5.3 PYONGYANG, c. 1800. (American Geographical Society Library, University of Wisconsin–Milwaukee Libraries: D469.1-d.P96 A-[18—?].)

ligraphy plays a central part not only in art and written language but also in mapmaking.

Other maps in the later Joseon carried on subjects and styles from the early Joseon. Representing relief from a bird's-eye view to lend it a more three-dimensional quality, for example, continues to give the maps the quality of a painted landscape. This is clearly seen in figure 5.3, a map of the city of Pyongyang made about 1800. The city is carefully delineated in precise brushstrokes, producing a work of art as well as an accurate map. This pictorial form of cartographic representation is common throughout the entire Joseon period.

The shapes and forces maps of the early Joseon also continue in the late Joseon. Figure 5.4 is a map of the suburbs of Seoul, probably produced in the last years of the Joseon dynasty. In many ways it is a modern map: it is printed on copper plates,

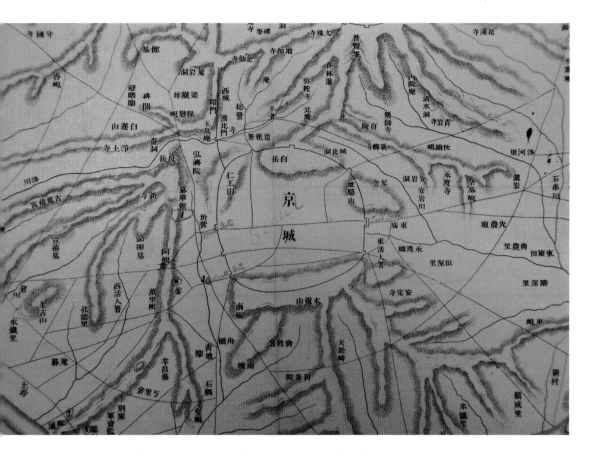

5.4   SEOUL, c. 1910. (Courtesy of the Library of Congress, Washington, DC: 7904 S478 .H3 Misc #151.)

hachures indicate relief, and the subject is the sprawling suburbs of the capital. But despite all the modern technology and symbols, the map is as much alive as maps of the region made under more traditional methods (see fig. 2.8). The link between the old and new maps embodies the enduring vitality of the shapes and forces map in Joseon cartography.

Maps in the late Joseon continue to deploy forms and styles developed in the early Joseon. Styles are copied down through the centuries, and there is a continuing tradition of pictorial cartography and of revealing the underlying geomantic forces in the landscape. However, the era also saw some new developments. The late Joseon is thus a time when old and new worldviews are brought into sharper relief. A. L. Mackay (1975), for example, considers the cartography of Kim Su-hong (1601–81), who was born in Kyongsang Province and became an influential secretary at the court, living and working in the early days of the late Joseon. He visited Beijing and brought back books and materials on Western science. His *Comprehensive Map of the Ancient and Modern World* was compiled in 1666. On the one hand it is old-fashioned, even for its time. It references Chinese material from 120 BCE and 180 CE and depicts the world map as a Buddhist mandala. However, while reflecting traditional beliefs in religious cosmology, the map also hints at newer systems of surveying, containing old perspectives as well as new ways of seeing the world. The cartography of the late Joseon embodies this tension. I will explore this creative tension in three of the most important cartographic forms: world maps, the atlas form, and new, more accurate national mapping.

CHEONHADO

Cheonhado can be translated as "map of the world beneath the heavens." *Cheonhado* maps depict a circular world centered on East Asia, with the rest of the world on the margins. These maps emerged from about the seventeenth to the nineteenth century. They most often appear as the first map in atlases. Figure 5.5 presents an example from about 1700. At its center is an inner continent consisting of China, Korea, Asia, Arabia, and Africa, surrounded by an inner sea then an outer ring of land. This highly stylized type of world map comes in a number of forms. Another map made about 1800, now in the British Library, notes China as the "central field." Moving farther from this central pivot, description becomes more imaginary than accurate, with places marked as "land of hairy people," "land of one-eyed people," and "land of refined ladies."

The *cheonhado* is a distinctive style of Korean map found in many atlases of the late Joseon. The focal point of the inner continent of these maps is Mount Kunlun in present-day Tibet, leading a number of scholars to posit a Tibetan Buddhist origin for these Korean world maps. Nakamura (1947), for example, argues that they draw on Buddhist cosmography from Sino-Tibetan sources and on Chinese maps no older than the eleventh century. The *cheonhado* maps, according to Nakamura, are in essence Buddhist mandalas as transmitted through the Chinese. There are a number of problems with this interpretation. Buddhism was discouraged in the early Joseon and so is unlikely to have survived unchanged through the formal encouragement of Neo-Confucianism. Ledyard (1994) proposes an alternative interpretation, suggesting that the continental outline in the 1402 *Gangnido* is the basis for subsequent changes in continental depiction. The shape of the continents in the *cheonhado* is the playful result of successive reorientations of the continents first outlined in the *Gangnido*. A more comprehensive interpretation is proposed

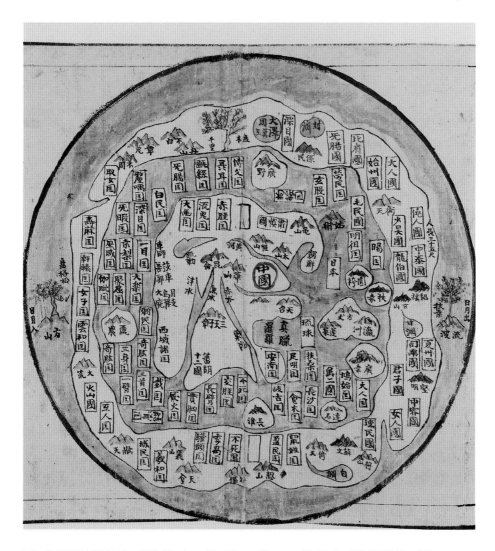

5.5 *CHEONHADO* MAP, c. 1700. (Courtesy of the Library of Congress, Washington, DC: G3200 17— .C5.)

by Oh (2008), who draws together a number of strands. The first is that these maps firmly reflect the Sinocentric view of round sky, square earth. The circularity of the maps represents the heavens, not the earth. In many *cheonhado* maps, star constellations are shown outside the circular border, reinforcing the conclusion that the circularity represents the celestial. Look again at figure 5.5. Inside the circular heavens the squareness of the earth is more obvious. The second strand is that the maps draw on much older Chinese scholarship. The names on the map recall those from much earlier Chinese texts, such as the *Shan hai jing* (Collection of the Mountains and Seas), written about 300 BCE. There are also echoes of the work of Zou Yan (305–240 BCE). These earlier writings were invoked because the contact with the West had undermined the more standard Confucian texts, with their limited knowledge of a wider world. In other words, the *cheonhado* maps were the result of contact with the West as interpreted by a Sinocentric worldview and informed by early Chinese geographical writings. To add to the rich mix, Oh suggests they contain elements of Daoism; the trees that mark the rising and setting sun in the east and west of the maps embody the sacred trees so important in Daoist beliefs.

The *cheonhado* maps represent less an old, unchanging text than a new text forged by the contact with the West and based on several contemporary and ancient sources, especially older Chinese texts. They are creative attempts to deal with the crisis of geographical knowledge initiated when traditional beliefs were ruptured by a radically new global geography. As in most rapid transformations, the responses adapt the contemporary to the historical, utilizing the tried and safe to deal with the troubling uncertainty of the new and the modern.

The *cheonhado* were not unchanging textual templates but were modified in successive representations. Figure 5.6, for example, is a world map with a grid. Made sometime in the eighteenth century, it has a graticule superimposed on the traditional world map. This example echoes the latitude and longitude scale used by the Sino-Jesuit cartographers. Again, China is marked as the "central field." Figure 5.7 is a world map from a manuscript atlas produced sometime in the middle of the nineteenth century. The continent outline is now more closely connected to standard accounts. An elongated North and South America can be discerned on the right side of the map, with the writing between them informing us that this is the "land of cannibals." This map draws on Ricci's world map and echoes the depictions of even earlier Western cartographers, such as Martin Waldseemüller's 1507 world map. This *cheonhado*, while keeping the basic circular form, has a more globally informed geographical representation.

ATLASES

Books were very restricted in the early Joseon, becoming more accessible only in the late Joseon. Atlases, especially small hand atlases, were one type of book that became more widely available. The paper was made by hand from mulberry bark. Korean atlases come in three main forms: woodblock, manuscript, and, later, cop-

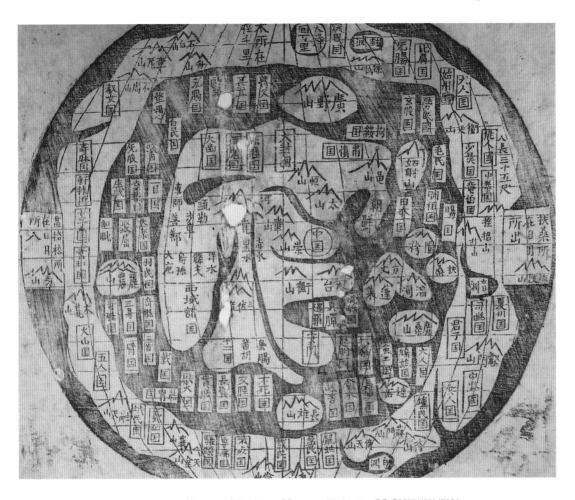

5.6 *CHEONHADO* MAP, c. 1760. (Courtesy of the Library of Congress, Washington, DC: G2330 Y651 1760.)

perplate. Manuscript maps were often copies of older maps. The printed atlases became common only in the second half of the nineteenth century. Atlases were an important part of the libraries of Korean scholars, often used in association with reading Chinese classics and standard Confucian texts. They were often more historical in their geographical information, showing outdated worldviews or anachronistic provincial boundaries. The atlases were less contemporary sources of knowledge than vehicles for situating texts and understanding the broad outlines of the country and the region; they represented general intellectual property rather than precise geographical information systems.

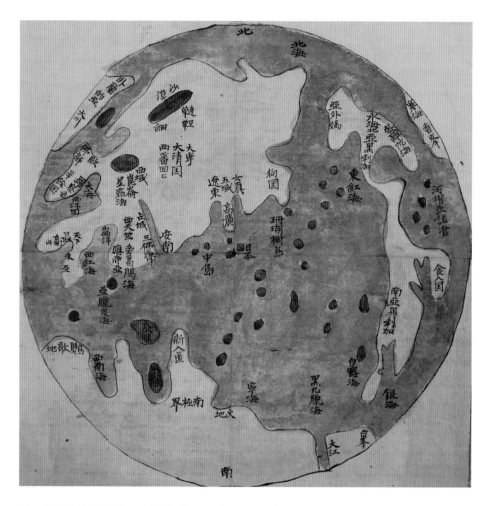

5.7 *CHEONHADO* MAP, c. mid-1800s. (American Geographical Society Library, University of Wisconsin–Milwaukee Libraries: AGS Rare At.469B-[1823–1869].)

Sometimes the atlases were scrolls where each map was pasted to the next and the atlas was unfurled a map at a time until it formed a line of maps. Figure 5.8 is a photograph of a nineteenth-century Korean scroll atlas that I laid out in the Map Reading Room of the Library of Congress in Washington, DC. This image highlights the "unfurling nature" of Joseon cartography. In this particular case, the maps have no color. Color was added later by hand. There was no standard color key.

The hand atlases have a common sequence. In the atlases with a larger geographic coverage, the first map, reading the suite of maps from right to left, is a

5.8 SCROLL ATLAS, nineteenth century. (Courtesy of the Library of Congress, Washington, DC: C2330 C53 18— Vault.)

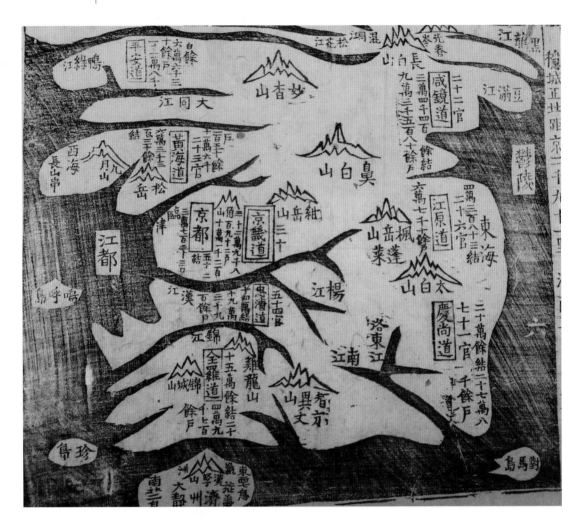

5.9 KOREA, c. 1760. (Courtesy of the Library of Congress, Washington, DC: C2330 Y651 1760.)

world map in the *cheonhado* style. Then the most common progression is for maps of China, Korea, and the separate provinces of Korea. Some atlases also contain maps of Japan and the Liu Chiu Islands. In these atlases Korea is depicted in three ways; first on the eastern edge of the map of China, then as a map on its own, and then as maps of its constituent provinces. Figure 5.9 is a map of Korea from a woodblock atlas made about 1760. Note the stark, almost geometric design, an example of the *sungnam* style.

It is interesting that these atlases with the widest geographical coverage depict only Korea and surrounding countries. There are no maps of Europe or the rest of the world. The world is portrayed through the stylized *cheonhado* and maps of China and Korea. By this time the geography of the outside world was better understood, yet these atlases still present a Korea firmly positioned within a tightly circumscribed Sinocentric world. They provided comfort more than contemporary geographical information, parallel to many European maps of the time that were only just beginning to depict Korea. While there are examples of cartographic encounters between East and West, Europe and Asia, practices of cartographic isolation continued. An early modern geographical idiom emerged slowly and fitfully.

There are also provincial atlases that contain maps of provinces as well as maps of each county in that province. A nineteenth-century atlas of Cholla Province, for example, contains fifty-four county maps. The map of the province highlights county seats in pink and watchtowers and forts in blue, and it shows temples as yellow oblongs. These more detailed county maps are in the pictorial tradition and show public buildings, temples, and official residences as well as accompanying text noting population size, acreage of arable land, and number of military reserves.

Provincial maps were made under the direction of local magistrates, who hired painters and local officials to draw them. Jeon (2008) discusses provincial maps in the *Haedong yeojido* (Atlas of Korea) made in the mid-eighteenth century and in the *1872 nyeon jibang jido* (Provincial Maps of Joseon in 1872). Both atlases were government-inspired works meant to produce a geographical information system for effective governance and administration. However, as Jeon demonstrates, the maps also contain two important features: a geomantic depiction of territory in the shapes and forces style, and a sophisticated use of different scales within the same map to highlight places of power. Core administrative areas and administrative buildings are shown in greater detail than the villages in more peripheral areas. Jeon describes the approach as "power holder–centered depiction of space." Indeed, most of Joseon cartography can be described in this manner.

There was growing cartographic literacy in the late Joseon as measured by the increasing number of both readers and texts. Besides traditional methods of cartographic representation such as mounting maps on fans, there were also the printing, publishing, and copying of atlases. These new cartographic forms circulated more widely through the population. Although there was innovation, there was also a heavy weight of inherited tradition as old maps were copied, reproduced, and repeated in diverse forms. But as maps diffused widely, there was no corresponding increase in global geographic awareness. The atlases tended to reinforce the position of Korea in a Sinocentric world. The *cheonhado* was a response to modernity that maintained this perspective. Loyalty to a China-centered worldview, national allegiance to the Joseon state, and nationalist sentiment regarding the land were highlighted in the depiction of Korea in the late Joseon maps and atlases, helping to establish and reinforce national consciousness rather than accurate global awareness.

NATIONAL MAPS AND NEW GEOGRAPHICAL UNDERSTANDINGS

In the late Joseon period, as part of the *silhak* movement, there were the beginnings of a more science-based geographic understanding and more accurate national and regional mappings. Chong Sanggi (1678–1752), for example, drew detailed provincial maps that constitute an early accurate representation of the country to a common scale. Chong Sanggi was one of the first cartographers to use a standard scale, at 1:400,000. Because the provinces varied in size, this meant that the pages containing provinces or groups of provinces might be of different sizes. Atlases in the Chong Sanggi tradition are often like complex works of origami, with sheets folded in intricate patterns.

Chong Sanggi's work was similar to the projects undertaken by imperial powers around the world to map and envision their territory. He was a contemporary of Louis XIV, the Qing emperor Kangxi, and Peter the Great, and his cartographic project was contemporary with national surveys in France, China, and Russia.

We know little about Chong Sanggi; he was not an official bureaucrat and practiced his scholarly pursuits in private. He wrote on a variety of subjects, including proposing to remove cash from the economy and trade only in cloth and rice. He believed in a simpler, purer Korea untouched by rampant commercialism. It was not until after his death that his maps attracted the government's attention. Those made about 1730 came to the notice of King Yongjo about 1757 and were sent off to the royal library for copying. They were copied numerous times by hand in atlases produced by the government and by private citizens and were the dominant form of national representation from the mid-eighteenth to the mid-nineteenth century.

Chong Sanggi's mapping drew on the Sino-Jesuit cartographic tradition that emphasized accuracy, maps made to a common scale, explicit use of the grid to organize and present geospatial data, and greater use of geodetic data. Contact with Sino-Jesuit mapmaking practices as well as those of indigenous cartographers led to new developments in Korean cartography. Chong Sanggi wrote that because of the distortion of previous maps, he decided to begin anew, mapping Korea at the scale of 1:400,000. His were some of the most accurate maps made of the nation to date, and in this regard his work undermines a reading of late Joseon cartography as simply copying earlier maps. It is more appropriate to consider late Joseon cartography as a site of creative tension between the heavy weight of tradition and

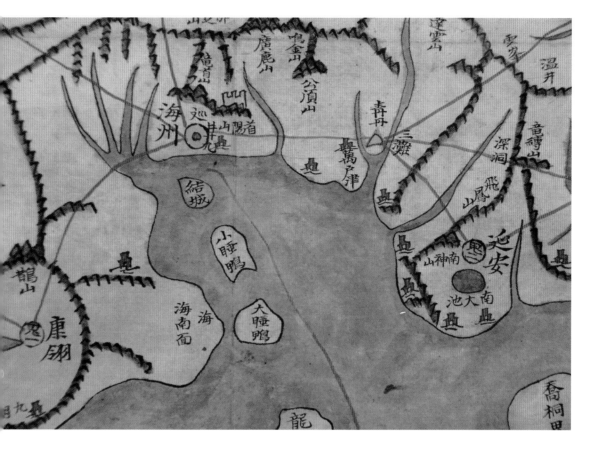

5.10 PART OF HWANGHAE PROVINCE, nineteenth century. (Courtesy of the Library of Congress, Washington, DC: G2330 T64 176 Vault.)

the promise of innovation, between the old and the new, with the established and the more experimental all contesting and informing the national discourse.

Figure 5.10 shows part of Hwanghae Province from a manuscript atlas made in the nineteenth century that is a copy of Chong Sanggi's 1730 map. This map is hand-colored: red indicates roads, yellow highlights county seats, signal towers are noted with red dots, and relief is shown by pictogram peaks colored green. The Chong Sanggi maps and their copies are in the shapes and forces style; mountain ranges and rivers alternate, with rivers running between the lines of mountains. The landscape is animated by this energetic depiction of mountains as arteries and rivers as veins.

Chong Sanggi style maps are found in a variety of manuscript forms. I have seen examples with a separate map for each of the provinces, others in which Kyonggi, Ch'ungch'ong, and Kangwon are placed on the same page, and yet others where only Kyonggi and Ch'ungch'ong are on the same page; some have Hamgyong on one page, while others have separate sheets for the northern and southern regions of this province.

The maps contain a great deal of information. Distances are given between towns, often with a starting point in Seoul, four classes of town are indicated, and garrison towns are highlighted.

To make his maps, Chong Sanggi climbed peaks to make sketches and used an odometer to measure distances and a plane table to estimate altitude. This mixture of precision alongside the shapes and forces style provides an important contact point between the local and the global, the premodern and the modern. Chong Sanggi's maps serve "as a bridge between the old cosmological and the new scientific cartography, and [are] thus an important milestone in the history of mapping in Korea" (Thrower and Kim 1967, 49).

The new cartographies were prompted by the contact with the Chinese and the Sino-Jesuit mapping tradition. From the early seventeenth century, envoys to China brought back Western-inspired maps and geographical understanding. In 1631 the emissary Jung Du-won returned from the Ming empire with Chinese translations of Western works on astronomy and geography. In 1644 Prince Sohyeon brought back from China some knowledge of Western science and scientific instruments such as clocks and telescopes. Contact with the Chinese and thus with Sino-Jesuit works brought Korea a new understanding of the world. This new knowledge was contested by some, as in the case of the *cheonhado*, but it was adopted and extended by others. Several scholars sought to use this new knowledge. Hong Dae-yong (1731–83) made scientific instruments to measure celestial bodies more accurately and thus laid the foundation for more accurate terrestrial mapping. Choi Hangi (1803–75)—we discussed him in chapter 1, see figure 1.3—made Korea's first globe as well as writing *Jigujeonyo* (Introduction to the Study of the Earth). Jeong Yak-yong (1762–1836) produced *Daedongjiji* (Korean Geography Book), which described the geography of the country and defined the latitude and longitude of every Korean city. In 1811 he also wrote a geographical text, *An Investigation of Our Nation's Territory*. Jeong Yak-yong had an interest in place-names but also a concern with the evolution of the nation's boundaries, leading to a narrative of national territory and territorial sovereignty. It is in this general context of adopting a more formalized science in association with practical learning and national sentiment that we can understand the great work of late Joseon cartography.

The late Joseon witnessed the culmination of national cartographic representation in the work of Kim Jeong-ho (1804–66), whose map of Korea is aptly described as the "grand summation of Korean cartography" (Ledyard 1994, 291). The map-maker and his work have entered national legend. His map was one of four in a special postage stamp edition produced in 2007 featuring old Korean maps. His name has even entered the heavens; in 2002 an asteroid was named after him. One romantic rendering paints him as a lowborn man who devotes himself to making an accurate map of the nation. He travels the country for years making careful measurements. He returns and, with the aid of his daughter, carves the woodblocks and prints a map so accurate that he is jailed for reproducing security-sensitive information that enemies of the state could use. He dies in jail. His mythology has a number of elements that explain its importance for contemporary Korea. His humble birth is an important trope for a modern Korea seeking a more democratic image and distance from the feudal past. His modest beginnings in association with his tremendous scientific achievement also have strong resonance with the aspirational nature of contemporary Korean society. Through hard work and dedication, someone not of the upper classes can achieve great things. And the subject of his work, a more scientific depiction of the country, feeds into modern nationalist sentiment that seeks to legitimate the present through connections with historical antecedents and revered cultural capital. Kim's legend fulfills a number of roles for contemporary Korea.

The scanty details of his life give creative space for these romantic renderings. We do know that he was born in humble circumstances and so does not appear in any official documents of the day. It is rumored that he was born in Hwanghae Province, though we cannot be sure. At some stage he moved to Seoul, where he made a living as a woodblock publisher. About 1834 he carved the woodblocks for a Western-style hemispheric map of the world. In the same year he also completed an atlas of Korea, *Map of the Blue Hills*. This atlas was not assembled by province but used two large sets of grid rectangles, twenty-nine of them from north to south, and twenty-two arranged east to west at a scale of 1:160,000. It is a very detailed rendering including administrative boundaries down to the subdistrict level and the post station network. It uses data from an 1828 government survey that contains the number of households, amount of arable land and grain production, military personnel and reserves, and distance from Seoul. Because Kim's maps draw from official government sources, they suggest a high degree of administrative support. The work contains a preface by the high official Choi Hangi, indicating that the maps had his sanction. He had access to centralized knowledge and backing from other high officials such as Choi Seong-hwan. One important data source was generated in 1791 when King Chongjo commanded the government observatory to generate more accurate geodetic data for various sites in Korea. These data provided a baseline for Kim's map.

We have a record of Kim's procedures. In a district sketch map he would lay out the river and mountains, draw concentric circles radiating from the administrative center, and plot various items including post stations, forts, schools, and shrines. These district maps were then compiled into a larger map.

Over the next three decades Kim undertook more survey work and refined his techniques in his 1861 masterpiece, *Daedong yeojido* (Map of the Great East [Korea]). It was based on his almost thirty years of survey work and study. The shape of the country was thus made more accurate, and distances more precise. On the map a distance counter was added to the roads every ten *ri*, approximately every four thousand meters. These counters provide a space-time map of the country; they are closer together in more mountainous regions, informing readers that the road is hilly and thus more arduous. Over thirteen thousand names were placed on the map, many of them corrected from the 1834 atlas. There is a wealth of detail using twenty-two separate symbols at the detailed scale of 1:160,000. Military sites are given special consideration. The map is arguably the best example of the shapes and forces style, with mountains as a unitary skeletal frame and rivers like blood vessels; the land is presented as a living body, with political and physical geographies combined into the national coherence of a living, breathing entity. Mountains are shown as continuously linked, sinuous sawtooth lines, the entire mountain chain system radiating out from Baekdu Mountain and back, suggesting both the underlying geomantic forces and the national cohesion. The map is wonderfully clean and simple, with few of the calligraphic flourishes that mark many Joseon maps. The atlas is distinctly modernist.

The *Daedong yeojido* consists of twenty-two folded documents that unfold to 20 centimeter by 30 centimeter cartographic transects that run east to west across the country. Together they form a huge map of the country 6.7 meters by 3.8 meters. The entrance to the Kyujanggak Archives at Seoul National University in Seoul displays a copy of this map nearly two stories high. Kim spent an entire year simply making the woodblocks. One version was printed in 1861, and twenty-two copies were produced. Another printing from the same blocks was made in 1864. A single woodblock sheet, a condensed version, was also published in 1861 (fig. 5.11).

Figure 5.12 is the island of Jeju as shown in an 1864 copy. This island, with an area of approximately two thousand square kilometers, is south of the Korean mainland. The map of this island is a typical Kim map and contains all the essential elements of his mapping of the entire country. The mountain system, in a geomantic design, is shown as a continuous set of lines emanating from the volcanic Halla Mountain. The map is accurate, full of information, yet clean and sharp; readers are not overwhelmed by too much information. The rivers are depicted as wavy lines while the roads are shown as straight. Using woodblocks makes it difficult to discriminate

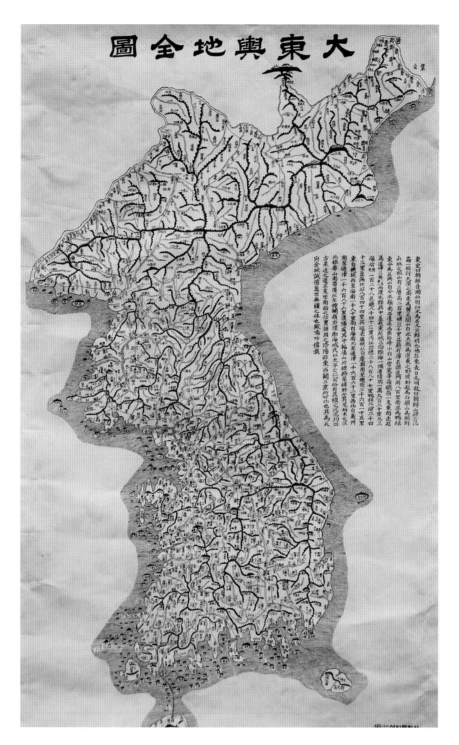

5.11 *DAEDONG YEOJIDO*, 1861. (Kyujanggak Archives, Seoul National University.)

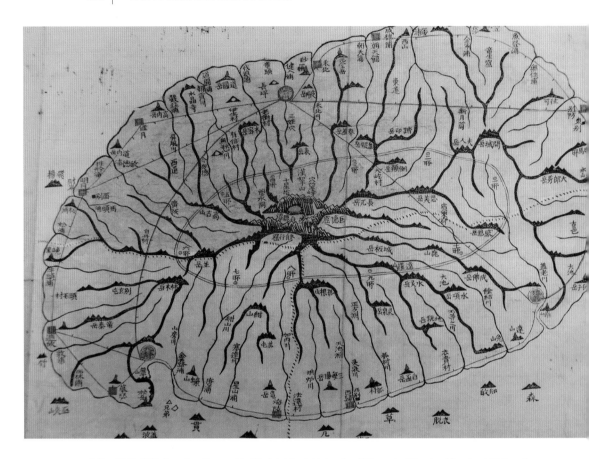

5.12 JEJU, 1864. (American Geographical Society Library, University of Wisconsin–Milwaukee Libraries: AGS Rare At. 469A-1861.)

between rivers and roads, so Kim used this wavy/straight distinction. The overall result is not only accurate but also aesthetically pleasing; the map shows lots of information in a precise yet beautifully understated manner, a model of cartographic clarity, accurate mapping, and scientific artistry. It is a successful hybrid map that embodies traditional geomancy and new science, old techniques and new methods.

Kim's maps are good examples of the new learning tied to policy issues. He wrote about his great work, "My map will be used to defeat the enemy and suppress violent mobs when the nation is troubled, and to carry out politics, govern every social affair and enforce economic policies in time of peace" (quoted in Chong 1973, 40).

The fragile political context is demonstrated by the quotation—Kim highlights internal unrest as well as external threat. As the nineteenth century progressed, social conflict deepened and outside menace increased. When Kim's map was printed, Korea was under pressure from foreign powers. The country skirmished with the French in 1866, the Americans in 1871, and the Japanese in 1875. The two hundred years of stability and seclusion in the late Joseon were coming rapidly to a bitter end. It is no accident that military sites received such detailed consideration in his map. The country was increasingly perceived as imperiled by internal social conflict and by interference from foreign powers. In the condensed one-sheet map of Korea also produced in 1861 (fig. 5.11), the text contains the flourish, "'Tis a storehouse of Heaven, a golden city! Truly, may it enjoy endless bliss for a hundred million myriad generations! Oh, how great it is!" There is an element of patriotic fervor, marked perhaps by an unspoken dread that the country may not enjoy endless bliss.

Kim's work is the signal achievement of late Joseon cartography. It combines accuracy with beauty, geomancy with geodesy, clarity with information, embodying practical learning as well as nationalist hopes. The map was produced at a critical juncture when two hundred years of Korean isolation was coming to an end, the five-hundred-year Joseon regime was in its last stages, and a modern world was bursting into life with constant change as its heartbeat.

NEW INSCRIPTIONS

The last decades of Joseon rule were marked by growing internal dissensions, increased external threat, and a forced entry into a wider world. The days of seclusion were ending as Chinese dominance drew to a close in the face of new geopolitical forces, including the rise of Japan as a military and expansionist power and the direct entry of the United States and European powers into the region. I will consider three map inscriptions that provide a window onto the last days of Joseon rule.

Attacks   Figure 5.13 is a detail from a larger map of Gangwha Province, in the northwest of present-day South Korea, made about 1870–80. At first glance it looks like a typical example of the pictorial maps so common throughout the entire Joseon period. It is at the same time an accurate map and a finely wrought landscape painting. But if you look closer you can see red lines emanating from buildings along the coast. I have highlighted this effect by rotating the picture so that east is at the top. The detail is of buildings along the shoreline around present-day Incheon, where the Han River enters the Yellow Sea, and the red lines represent cannon and rifle fire. The quiet elegance of the map is disturbed by the marks of war. It records a site where the Koreans first had an armed encounter with a

5.13 GANGWHA PROVINCE, 1870–80. (National Museum of Korea.)

Western power, then experienced subsequent attacks and defeats. Although I cannot be sure, I have a sense that these buildings and lines of fire were added to an existing map. They seem jammed in, not in scale with the rest of the map, as if they were put in later.

The Joseon kingdom maintained a policy of seclusion, but by late nineteenth century the Western powers and resurgent Japan were keen to break into Korean markets. This was the era when treaty ports were forced on China and Japan. Although contact between Korea and the West was severely limited, there was some missionary activity. Catholic missionaries traveled there from China, and by 1860 there were about eighteen thousand converts. In 1866 the Joseon regime rounded up the priests, most of them French, as well as native converts, and executed them. When news reached French forces in China they decided to launch a punitive mission, but they had no accurate geographical information. In September and October 1866 they made some navigational charts around Gangwha Island at the mouth of the Han River. Rather than attacking the heavily fortified capital of Seoul, on October 16 they occupied Gangwha Island with 170 marines. When they tried to land on the mainland, they were rebuffed. Facing an ever-increasing Korean defense buildup, the French soon sailed away. There is a cartographic postscript to this event. Ganghwa Island was an important annex of the Joseon Royal Library, and the French looted treasures that now form the basis of the Korean collection at the French Bibliothèque Nationale. Han (2008) traces the story of one such stolen treasure, a beautiful large hand-painted world map made between 1637 and 1644.

The French were repelled, but this did not deter other Western powers. Only five years after the French sailed away in defeat, the Americans arrived with five warships to support an American delegation seeking to open Korea to trade and treaties. As the United States fleet sailed through the Gangwha Strait leading to the mouth of the Han River, Korean forces fired on them. The rear admiral in charge wrote of cannons in rows placed on the hillside. The cannon fire of this encounter may account for the red lines we see in figure 5.13. The fusillade failed to sink any ships, and the Americans launched a punitive attack that killed almost 250 Koreans and captured five forts. The encounter strengthened the hand of the isolationists in the Joseon court. The Joseon government refused contact and reimposed an isolationist policy. Soon, however, the area was again subject to foreign penetration when in September 1875 a Japanese warship launched a small boat toward Gangwha Island. The Korean forces opened fire. The line of cannon fire shown on figure 5.13 was perhaps employed yet again. The next year a larger Japanese fleet forced the Joseon rulers to sign the Treaty of Gangwha creating the three treaty ports. The coastline depicted in figure 5.13 was the scene of foreign penetration first by the French, then by the Americans, and finally by the Japanese, forcing open the previously isolationist Joseon regime.

Treaty Ports   Figure 5.14 is a hand-colored woodblock map of Korea first printed perhaps sometime between 1857 and 1866 and distributed widely. It depicts the country at a uniform scale of 1:700,000, with the land floating in a sea of text. It is a document celebrating the nation-state. There is a copy of this nationalistic map in the American Geographical Society (AGS) Map Library in Milwaukee. Americans living in Korea in the late nineteenth or very early twentieth century most likely procured it. The AGS copy is distinctive because it is marked with comments in English. The handwritten English key shows a red anchor symbol beside the three ports of Busan, Incheon, and Wonsan, the three treaty ports imposed on Korea by Japan in 1876. Figure 5.15 highlights the port of Busan.

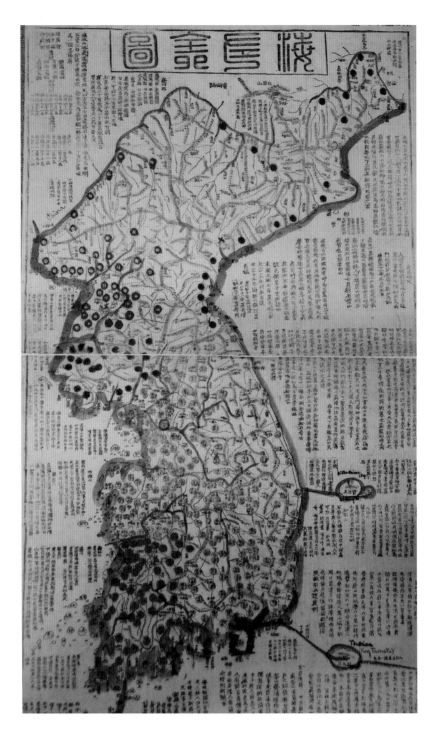

5.14 KOREA, 1857–66. (American Geographical Society Library, University of Wisconsin–Milwaukee Libraries: Rare 469-A 1857–1866.)

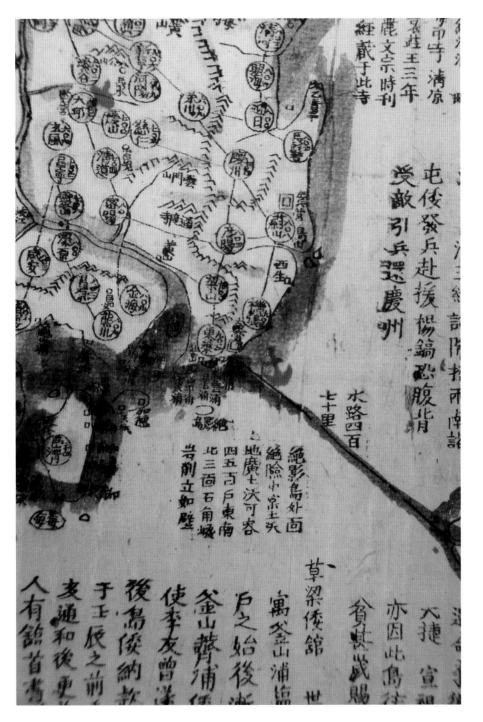

5.15 DETAIL FROM MAP OF KOREA, 1857–66. (American Geographical Society Library, University of Wisconsin–Milwaukee Libraries: Rare 469-A 1857–1866.)

Treaty ports, where foreigners were allowed access to markets, residence rights, and freedom from local authority, signified a loss of national sovereignty and a forcible entry into global trading patterns. Treaty ports were initially forced on China and later Japan by the Western colonial powers. The British imposed the first on China after they successfully defeated the Qing empire in the first Opium War of 1839–42. Sixteen Chinese treaty ports were established by 1860, and more than fifty by the end of the century. Foreigners could reside there exempt from internal tariffs and not subject to the national laws and regulations. In 1854 United States warships under the command of Commodore Matthew Perry "opened" the Japanese port of Shimoda, the same year a Russian fleet opened up Nagasaki. Forcing treaty ports became the preferred form of neocolonial dominance.

Beginning in 1866 the French, Americans, and Russians attempted to force treaty ports on Korea but were repelled. The Japanese were more successful after an incident in 1875 when a Japanese ship attempting to land a small boat on Korean soil was fired on. A Japanese fleet returned the next year in an ominous echo of Perry's actions against Japan. The Japanese fleet, consisting of two warships, three troop transports, and a ship used as a floating embassy, forced the Treaty of Ganghwa that opened up the three treaty ports and pried Korea away from its tributary status with China.

Treaty ports imposed by outside powers signified the end of economic and political sovereignty. They also transformed local places. Choe (2005) describes the effects of treaty port status on Incheon as a tiny village was transformed into a spreading city with areas set aside for Japanese, Chinese, and Westerners, its population growing to almost fifty thousand by 1910 and the number of Japanese increasing to about fifteen thousand. Incheon and the other treaty ports were a beachhead for Japanese control of Korea and the entry point for global penetration. Figure 5.16 is an undated manuscript map of Incheon, made soon after the 1876 treaty, that shows the demarcated areas of Japanese and Chinese settlement close to the foreshore.

The Foreign Presence   Previously I noted a pictorial map of the city of Pyongyang made sometime in the nineteenth century (fig. 5.3). It also is part of the AGS Library collection, a gift of Americans who lived in Korea. And it too contains a later inscription. Various features are noted in red ink. The unknown writer has marked, in English script, the main gateways, the residences of the governor and the judge, as well as the presence of Christians. Figure 5.17 highlights the Presbyterian Church next to the East Gate. In another part of the map the property of the church is highlighted. The marking on this map shows the incursion of foreign

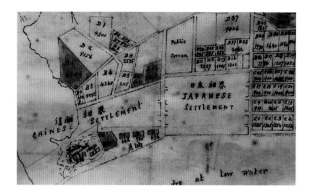

5.16 INCHEON, late nineteenth century. (National Museum of Korea.)

missionaries into Pyongyang in particular, but also into Korea in general. These markings were probably made in the early twentieth century, and they represent contemporary missionary activity in Korea.

Whereas Catholic and especially French Catholic missionaries had been coming to Korea since the eighteenth century, Protestant missionaries such as the Presbyterians did not arrive until 1885. They were predominantly from the United States. Christianity was outlawed during much of the late Joseon, but King Kojong (r. 1863–1907) supported missionaries in order to modernize the country and to secure American assistance. Treaties signed with various countries—the United States in 1882, Britain in 1883, and France in 1886—extended the missionary reach. In the 1882 treaty, missionaries were originally limited to treaty ports. In 1883 they were allowed to travel anywhere in Korea and to live up to thirty-three miles from a port, but they could erect places of worship only in treaty ports. An 1886 treaty effectively allowed Christian missionaries to work throughout the country and gave them extraterritorial rights. American missionaries, for example, were subject only to United States law, thus gaining enormous independence and semidiplomatic status. And they were allowed to own property. American missionaries increased in number and importance (Young 2003). They built schools and hospitals and were able to convert many Koreans to Christianity. Pyongyang was an important site of Christian endeavor, so much so that it was referred to as the Jerusalem of the East (Baker 2003). The markings in figure 5.17 indicate the more pronounced foreign missionary presence in the last decades of the Joseon era. The missionaries also played an important role in the resistance to subsequent Japanese occupation of the country. Some scholars expand the discussion to theorize missionaries as agents of a modern scientific worldview (Dunch 2002). Here I have simply shown

5.17 DETAIL OF CITY OF PYONGYANG, c. 1800. (American Geographical Society Library, University of Wisconsin–Milwaukee Libraries: D469.1-d.P96 A-[18—?].)

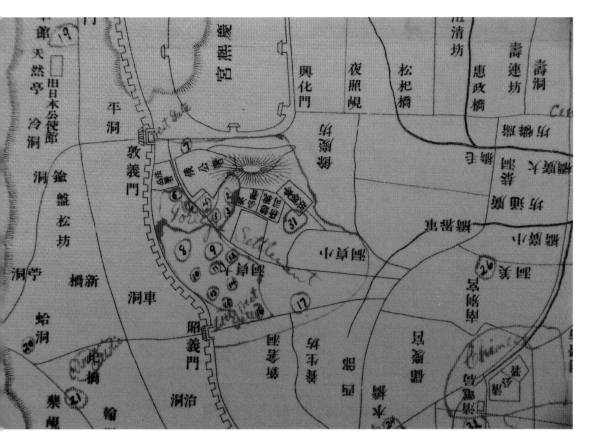

5.18   SEOUL, c. 1900. (Courtesy of the Library of Congress, Washington, DC: G7904.S478 18— .K9 Vault.)

how their presence is recorded on a Korean map—perhaps a useful metaphor for their effect on local culture. Their more precise role in the construction of a global space needs more careful analysis.

In the last decades of the nineteenth century and the first decade of the twentieth, Korea was forcibly opened up to the outside world. An indication of a foreign presence is revealed in English inscriptions on another map. The city of Seoul was a popular topic of late Joseon cartography. In the late nineteenth century numerous maps—manuscript maps, woodblock maps, and maps printed on copper plates—circulated widely. As the city grew, companion maps also were produced—a detailed map of the city proper and another of the wider city region. Figure 5.18 is a detail from a copper-engraving map of the city made on the eve of Japanese occupation. Rivers and roads appear to be the result of ground survey. The walls around the city are clearly visible. Someone, most likely American missionaries, has indicated in English text specific features such as the East Gate, the Palace, the Old Palace, and the areas of foreign settlement, including the Chinese area and the Japanese area in the south and southwest areas of the city. The top right corner of the map also has a list of numbers indicating residences, churches, missions, hospitals, and schools. Part of the key reads:

4 Mr. Miller
5 Mr. Gifford
6 Boys' School
7 Presbyterian Church

Figure 5.18 highlights the Chinese and foreign settlements areas clustered around the West Gate; 8 is a Methodist girls' school, 29 is a Methodist chapel, and 31 is an English church mission. A companion map of the suburban region within a three-mile radius also shows chapels and missionary residences. The inscriptions on the map highlight the marked foreign penetration into the city. English inscriptions on a Korean map represent a new era, the end of Joseon seclusion and the beginnings of a contemporary Korea.

REPRESENTING KOREA
IN THE MODERN ERA

6

THE COLONIAL GRID

Figure 6.1 is a Japanese grid map. Made in 1921, it captures Korea in a series of cartographic meshes. The map represents, embodies, and symbolizes the Japanese domination of Korea from 1910 to 1945.

The straight-line grid is one of the distinctive designs of a high modernist ideology, a particular form of modernism that James Scott (1999) describes as a belief in progress and a powerful state's desire to make society conform to a rational plan. This "seeing like a state," the title of Scott's book's, is all the more forceful because the muscular state in this case was Japan, exerting its dominance over Korea in a context of colonialism. This grid is especially evocative of imperial control and imposed order. It provides the necessary cartographic net to collect and display information as well as the imperial signature that signifies rational order and imperial control. Control justified by order, order reinforced by control. The grid both symbolizes colonial domination and signals Korea's forced incorporation into the production of global space.

Korean-Japanese relations varied over the centuries, at times violent, at times peaceful. A joint Korean-Mongol force invaded Japan in the thirteenth century. And in the very late sixteenth century there were two Japanese invasions of the Korean peninsula that caused enormous damage. All the government's precious world maps were lost in the invasions. Han (2008) argues that the beautiful hand-painted large world map made between 1637 and 1644, now in the Bibliothèque Nationale,

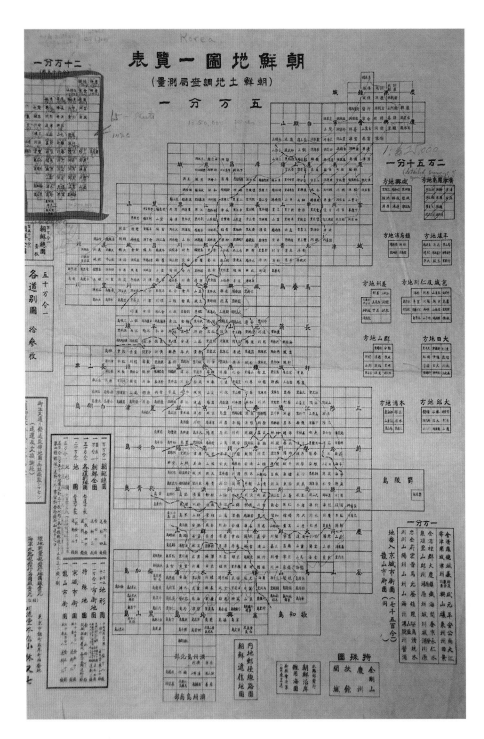

6.1 JAPANESE GRID MAP OF KOREA, 1921. (American Geographical Society Library, University of Wisconsin–Milwaukee Libraries: 469 D-1921.)

was made to replace the world maps either looted or destroyed. Japanese pirates also caused continual problems for Korean coastal communities. But from the early seventeenth century until the late nineteenth, Japan and Korea had relatively amicable relations. All this was to change with the decline of the Joseon regime and the rise of a more militaristic and expansionist Japan.

In the Meiji period (1868–1912) Japan followed a policy of economic modernization and military buildup. The humiliation of the 1854 Treaty of Kanagawa, which forcibly opened up Japan with the power of Perry's gunships, revealed the costs of seclusion, economic backwardness, and lack of military might. In its wake, Japan hired thousands of foreign experts and quickly adopted modern technology as it sought to modernize, industrialize, increase its military strength, and improve its global position. It also embarked on mapping nearby states. Figure 6.2 is a map of Korea with three insets of harbor maps made by the Japanese in 1875. A Japanese warship secretly surveyed Korean coastal waters, and the three insets represent clandestine harbor surveys, preemptive cartographies ready for possible incursions. Figure 6.3 shows the one for the Tai-Tong River; the English words and numbers are a later addition. This map was made just before Japan imposed the 1876 Treaty of Ganghwa on Korea, forcibly opening the country to foreign trade and prefacing the start of Japanese domination. In 1876 Japan forced the Joseon rulers to open three treaty ports. This carefully delineated Japanese map of Korea, with the harbors of the three future treaty ports meticulously charted, is one of the earliest documents that indicates the beginning of the end for the Joseon regime and the start of the troubles that lay ahead for Korea.

The Sino-Japanese War of 1894–95 was essentially a struggle over Korea. The general background was the decline of the Qing empire, the increasing weakness of the Joseon dynasty, and the growing strength of Japan. The specific context was growing conflict and rising tension between China and Japan over control of Korea. In 1882, in the wake of droughts, famine, and a fiscal crisis, riots in Seoul led to attacks on the Japanese legation. Japan sent warships and troops to Seoul, and China countered with 4,500 troops. A treaty signed in 1882 reduced the immediate conflict, but the underlying tensions did not disappear. In 1884 a pro-Japanese coup was succeeded by a pro-Chinese coup, and tensions again increased between Japan and China, with yet another treaty signed in 1885. But the antagonism continued. In 1894 a peasant rebellion led the Joseon to ask for Chinese troops—almost 2,500 were sent. The Japanese countered with 8,000 troops, who captured Seoul in June 1894, established a pro-Japanese faction, and declared Korea independent from Chinese control. War between China and Japan was officially declared on August 1, 1894. The Japanese army defeated Chinese troops defending Pyongyang, and the Japanese navy destroyed Chinese warships at the mouth of the Yalu River.

6.2 KOREA, 1875. (American Geographical Society Library, University of Wisconsin–Milwaukee Libraries: 469 D-1875.)

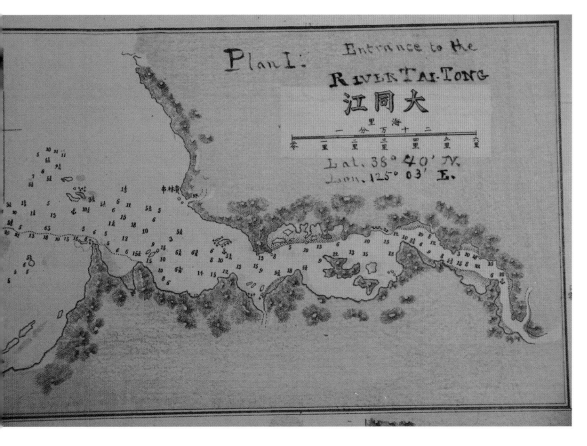

6.3 INSET FROM MAP OF KOREA, 1875. (American Geographical Society Library, University of Wisconsin–Milwaukee Libraries: 469 D-1875.)

The Japanese army marched into Manchuria, and the Japanese victory was officially recorded in the Treaty of Shimonoseki on April 17, 1895, when Korea was declared independent from China. The treaty also allowed the Japanese to sail ships in the Yangtze River and operate manufacturing plants in treaty ports. The treaty ceded Taiwan to Japan and forced China to open up four new treaty ports and pay substantial reparations to Japan. Korea's traditional relationship with China was destroyed, and Japan was now dominant in the region.

With this victory, Japan became a major power in the Far East. Eager to maintain its control over Korea, it did not hesitate to take on another regional power, the Russian empire. The Russo-Japanese War of 1904–5 was in part a struggle over Korea as well as Manchuria. With yet another victory in this war, Japan secured global power status and saw its effective control over Korea recognized by the international community. Korea was now part of the Japanese sphere of influence. It became a Japanese protectorate in 1905 and came under formal Japanese control with annexation in 1910.

The Japanese takeover of Korea occurred at a time before the international community recognized human rights. Geopolitical considerations trumped people. The very idea of human rights or national self-determination had yet to appear in diplomatic considerations. There was very little comment and no action on the Japanese takeover of Korea. The United States gave Japan a free hand despite having signed a treaty with Korea in 1882 and despite the promptings of some American Protestant missionaries critical of the harsh Japanese rule (Agata 2005). The places inscribed on the maps of Korea shown in figures 5.17 and 5.18 were often centers of resistance to the worst excesses of Japanese power.

Under Japanese rule, Korea became a colonial society. Its economy was restructured to fit Japan's need for rice, for raw materials such as timber, for sites of heavy industry, and for cheap labor. The formerly closed Korean economy was reoriented to the needs of the Japanese economy. A form of cultural imperialism also sought to replace a Korean consciousness with a Japanese view of the world. There was collaboration as well as resistance, accommodation as well as uprising. Some Koreans worked with and for the Japanese, while others resisted. Guerrilla campaigns were fought, a provisional government was established in Shanghai, and anti-Japanese protests and demonstrations took place. In the largest, almost two million protesters demonstrated in 1919. The form of Japanese control also shifted over the period. In the wake of the 1919 demonstrations, the Japanese relaxed some of their controls, but they reimposed them after a student uprising in 1931.

Although some revisionist historians argue for the benefits of Japanese rule, suggesting that it laid the foundation of the capitalist export economy that propelled South Korea from the 1960s onward into an important economic power, the association was always unequal and unfair. Japan imposed its will on its neighbor in a one-sided suzerainty.

Japanese colonial domination took many forms. Mapping is just one. The country was surveyed as a form of geographical intelligence gathering, especially to take an inventory of the nation's stock of land and resources. The Japanese undertook a series of cartographic surveys of Korea, some predating formal annexation in 1910. Japan carried out hydrological surveys along the Korean coast throughout the latter half of the nineteenth century. In 1894, at the outbreak of the Sino-Japanese War, the Japanese established a provisional department of Land Survey. Surveyors were dispatched to make military maps of Korea, but they had to do so in secret because they faced popular opposition. The resistance caused the department to be dissolved in 1896. It was reestablished in 1904, however, and Japanese

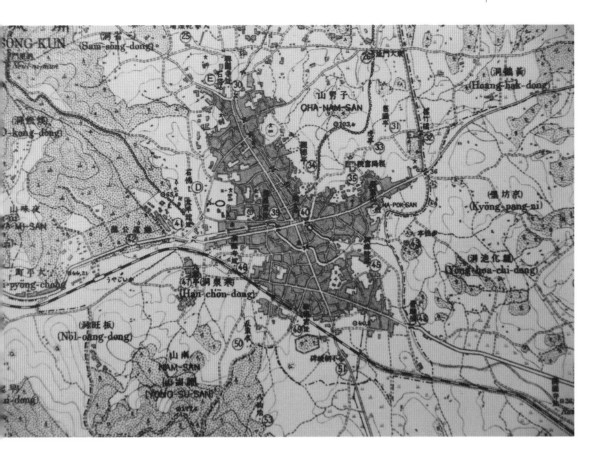

6.4 1:25,000 MAP, 1917. (American Geographical Society Library, University of Wisconsin–Milwaukee Libraries: 469 D-1917.)

mapmakers surveyed Korea from 1894 to 1908. These secret maps formed the basis of later maps made by the Japanese colonial government.

Figure 6.4 is a map made at the scale of 1:25,000. It is officially dated 1917 and was mapped by the Japanese General Staff, although it probably drew on the earlier secret maps. It is one of the maps made of tourist areas and had an English index and romanized names as well as Japanese script.

Korea was mapped as a form of imperial control. The Japanese published maps of Korea at a variety of scales, including 723 sheets at 1:50,000; 65 sheets at 1:200,000; 13 provincial maps at 1:500,000; 98 maps of Korean cities at 1:25,000; and 41 urban maps at 1:10,000. Figure 6.5 shows part of a 1916 Japanese map of Korea made at the scale of 1:200,000. Land use maps were also prepared at 1:500,000.

Detailed mapping was also an essential part of the nationalization of land and

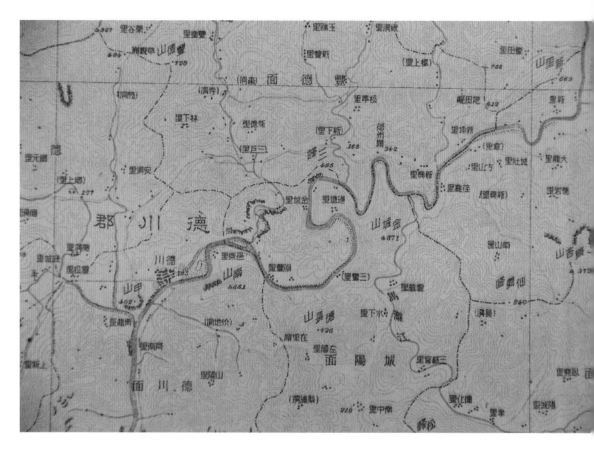

6.5 1:200,000 MAP, 1916. (American Geographical Society Library, University of Wisconsin–Milwaukee Libraries: 469 D-1916.)

of royal forests. Land was a major source of wealth and status in Korea, since land-ownership was the traditional power base of the *yangban* class and taxes on land were the major source of government revenue. Landownership was unevenly distributed, with a small minority having vast holdings. The Japanese colonial government wanted both to tax land and to control it. The Land Survey Bureau completed major surveys between 1910 and 1918, recording all plots of land and classifying them according to type, productivity, and ownership. The Japanese authorities required every owner to register his land claims. Many of the smaller owners and tenants could not document their ownership, but the large landowners generally kept their land. Almost 40 percent of arable land was taken over by the colonial government, and much of it was assigned to Japanese development companies at low prices.

6.6 MAP KEY, 1917. (American Geographical Society Library, University of Wisconsin–Milwaukee Libraries: 469 D-1917.)

The mappings gave a firmer basis for Japanese control over the land and resource base of Korea. Figure 6.6 shows the key employed in a 1:25,000 map. Notice how the land is carefully calibrated by type, indicating ginseng fields, mulberry fields, and paddy fields. When there were food shortages in Japan, food was sent from Korea.

There was also a toponymic colonialism in the new Japanese mapping of Korea, part of broader suppression of the Korean language, especially name usage. In 1939 Koreans were encouraged to change their surnames to Japanese forms. This linguistic colonialism was also evident in maps. Japanese place-names appeared in maps of Korea from 1900 onward. Maps of Seoul showed the name Keijo or Kyongsong. From 1914 onward a new Japanese naming system was applied to maps (Short and Lee 2010). Four methods were used. First, new administrative place-names were adopted after annexation that added Japanese generic names such as Tong, Jeong, and Jeongmok or directly copied Japanese names such as Bonjeong, Gilyajeong, Hwang Gumjeong, and Taepyong Tong. Second, roads and natural features were renamed in Japanese style. Third, besides Japanese inscriptions there were erasures of Korean. In Seoul, for example, 661 administrative districts with Korean names were replaced with 186 districts with Japanese names. Finally, larger regional features were also renamed. The sea between Japan and Korea had multiple names through the centuries; the Koreans predominantly called it East Sea. After annexation as part of the new wider Japanese empire, it was named Sea of Japan. More of this later when we deal with contemporary "cartroversies."

THE POSTCOLONIAL, POSTWAR WORLD

The defeat of Japan by the Allied powers in World War II broke the chains of Japanese colonialism in Korea. As a defeated power, Japan had to give up its imperial ambitions. Korea, however, was now caught up in a wider geopolitical struggle between East and West. The Soviet Union and the United States divided the Korean peninsula into separate zones of occupation. Figure 6.7 is a 1946 American State Department map that outlines the areas of occupation overlaid on the provincial boundaries; the peninsula was simply divided in half along the thirty-eighth parallel. The initial plan, devised in the immediate postwar period, was to place the country under the trusteeship of the United Nations. But competing national interests, eager to grab their chance, led to partition. In 1948 the Republic of Korea was established in the south under the leadership of Syngman Rhee, while Kim Il-sung led a communist system in the north.

The rising tensions soon gave way to all-out war. Each of the two leaders wanted to reunify the country under one political system with himself as head. Rhee, who had established a Korean government in exile, wanted an anticommunist system with himself as leader. Kim Il-sung, who had fought in the guerrilla campaign against the Japanese, wanted to create a communist system throughout the peninsula. After border sparring by both sides, North Korean troops crossed the border on June 25, 1950. Figure 6.8 shows the cover of a map of Korea made in 1950 in which soldiers and tanks figure prominently. The North Koreans quickly overwhelmed the weak South Korean forces and rapidly moved south. The United States and the United Nations now sent troops to defeat the North Koreans and pushed the battle lines all the way back north past the thirty-eighth parallel and deep into North Korean territory. The American advance prompted the People's Republic of China to send troops. Then the North Koreans and Chinese successfully pushed the United States and United Nations forces southward. By 1951 the competing forces were almost back to the thirty-eighth parallel. The war, which left the border very much where it was before, caused about 2.2 million military

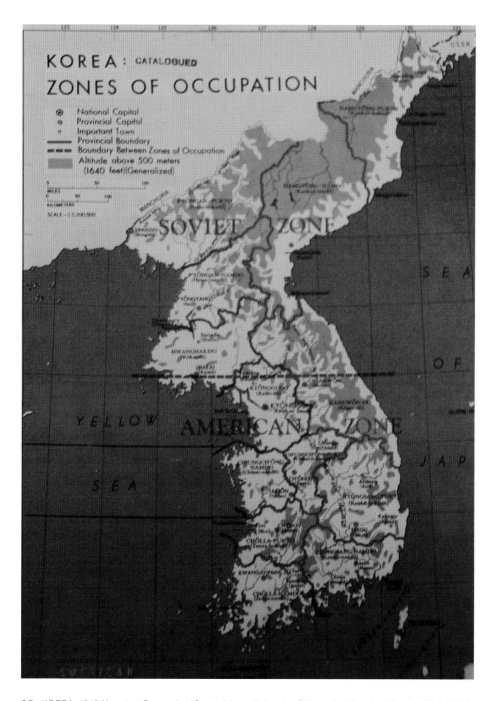

6.7 KOREA, 1946. (American Geographical Society Library, University of Wisconsin–Milwaukee Libraries: 469 B-1946.)

6.8 MAP COVER, 1950. (American Geographical Society Library, University of Wisconsin–Milwaukee Libraries: 469 1950.)

casualties and deaths and left between 2 million and 2.5 million civilians dead or wounded. Armistice negotiations began in 1951, and an agreement was reached in July 1953. No peace treaty was signed. The 1953 Armistice was an agreement between military commanders of the United Nations, North Korean, and Chinese forces to stop fighting, exchange prisoners of war, and establish a demarcation line

6.9 BORDER REGION, CIA, 1972. (American Geographical Society Library, University of Wisconsin–Milwaukee Libraries: 469 D-1972.)

with a Demilitarized Zone stretching two kilometers on each side. A map was attached to the agreement outlining the boundary. The border was established roughly at the thirty-eighth parallel. The signatories saw the document as a prelude to a formal peace agreement. However, South Korea never signed the Armistice Agreement, and a peace treaty has still to be signed. The military cease-fire froze relations on a permanent war footing. In principle a formal peace treaty that could provide some political resolution to the conflict remains a tantalizing possibility, but in practice the continuing uncertain and unresolved arrangement constitutes a dangerous situation. Since the two Koreas have yet to sign any such peace treaty, national sovereignty remains a clouded, contested issue. The cartographic consequences of this liminal arrangement will be explored in the later discussion of "cartroversies."

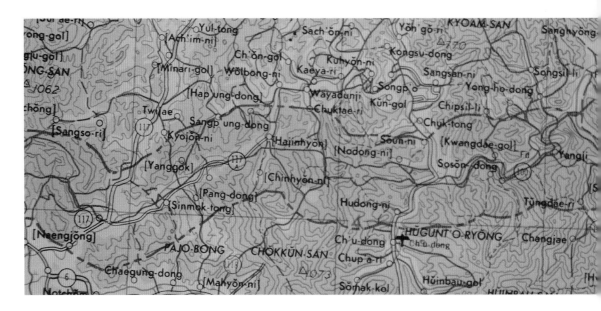

6.10 DEMILITARIZED ZONE, 1969. (Courtesy of the Library of Congress, Washington, DC: G7909 .F2 1969 U5.)

Figure 6.9, a map made by the CIA in 1972, depicts the border between the North and the South. Notice how the military airfields and landing strips are shown in North Korea. The border, while it crosses the thirty-eighth parallel, does not strictly follow it. The demarcation on the map represents the continuing divide between the two Koreas. Figure 6.10 is a more detailed map of the Demilitarized Zone made by the CIA at the scale of 1:250,000. This "temporary" armistice line has become a fixed element in the Korean landscape, a narrow but deep scar dividing North from South.

7 REPRESENTING THE NEW COUNTRY

Since 1953 Korea has been split into two separate political regimes. The one in the North, ostensibly a communist system, is a closed totalitarian society having limited contact with the outside world, a fragile economy, and a unique foreign policy that makes it a pariah in the global community of nations. South of the border lies South Korea, which has become a more democratic society and an important player in the global economy, increasingly inserted into global flows of capital, ideas, and people. In this chapter I will look at how the two countries are represented by others and by themselves.

REPRESENTING NORTH KOREA

North Korea occupied an important position in the immediate postwar world. It was on the front line between the East and West, precariously positioned on the sharpest edge of the bipolar world. This unique position engendered multiple mappings of the country by the competing superpowers.

Figure 7.1 is a Russian map of Pyongyang, the capital of North Korea, originally made about 1982. It was released only when the old Soviet archives were opened up after the fall of communism in 1989. The Cyrillic script is an indication of the geopolitical space that Korea occupied for much of the postwar world. It was on

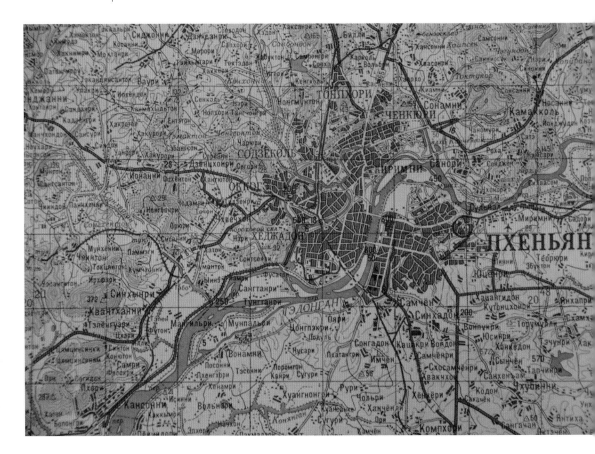

7.1 PYONGYANG, 1982. (American Geographical Society Library, University of Wisconsin–Milwaukee Libraries: 469-d.896 A-1982.)

the front line of the formal cold war: south of the Demilitarized Zone (DMZ) was South Korea, with between 28,000 and 40,000 United States troops in over twenty bases and camps.

North Korea occupied a difficult position during much of the cold war. It shared borders with China and the Soviet Union. China came to the aid of the North Koreans during the Korean War, with the Chinese pushing back American, South Korean, and United Nations military forces. However, the Sino-Soviet split, which lasted from 1961 to the 1980s, placed North Korea in an uncomfortable spot, geographically close to the two warring countries and strategically situated between them. North Korea tended more toward nonalignment, similar to Yugoslavia. Both China and the Soviets wooed North Korea, and in turn Kim Il-sung sought

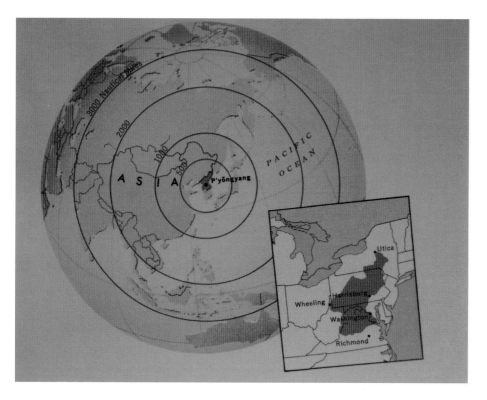

7.2 NORTH KOREA, CIA, 1969. (Courtesy of the Library of Congress, Washington, DC.)

to decouple economic and military aid from explicit support for either side. While some of the North Korean domestic polices emulated those of China under Mao Tse-tung, the North Koreans maintained relations with both countries. The country's geographic position necessitated a pragmatic response of careful diplomacy reflecting the country's vulnerable political geography.

The country was extensively mapped by both superpowers. Figure 7.2 is a detail from a CIA map made in 1969 that situates North Korea in absolute space and also compares its size with that of the United States. It is perhaps not incidental that the country's superimposed outline touches on Washington, DC, perhaps suggesting its importance to American geopolitical strategy in the cold war. Since 1989 the cold war has ended, but not in the Korean peninsula. No formal peace treaty has yet been signed, and North Korea continues to remain a thorny problem for the United States as guarantor of South Korea's security. There has been extensive military and civilian mapping by the West of North Korea's economic, political, and military infrastructure, including the secret United States military mappings as well as more open source information. The location of missile sites, air bases, and grain storage depots are now routinely available on the Internet.

Communist states have crumbled around the world. Even as remaining communist regimes such as China and Cuba adopt more market-oriented systems, North Korea continues to operate a command economy that is by most accounts failing miserably to meet the most basic standard of providing enough food for all its citizens. North Korea has slid into greater isolation, a rerun of the Hermit Kingdom experience of the later Joseon era. Yet political isolation in the modern world does not bring cartographic exclusion or silence. With the advent of Google Earth, even the most closed regime cannot escape the panopticon of the ever-present satellites in the sky. A website called North Korean Economy Watch (http://www.nkeconwatch.com/north-korea-uncovered-google-earth/) now uses images from Google Earth along with ground truth observations to create satellite imagery maps of North Korea. The website, according to its founder, is a form of democratized intelligence that allows viewers to see a satellite map of the country, including the more secretive elements such as compounds of the elites, prison camps, and new markets springing up outside cities. One map shows just one of the many palatial compounds used by the North Korean leaders. Even the most closed societies inhabit a shared planet. Hermits are watched and mapped despite their demand for seclusion. These newer forms of the production of global space allow ordinary citizens with a computer and Internet connection access to the human geography of even the most secretive regimes.

REPRESENTING SOUTH KOREA

Since the ending of the Korean War, South Korea has undergone massive changes, including growing resistance to authoritarian government, a greater democratization, and phenomenal economic growth. It is now the world's fifteenth largest economy and the eleventh largest trading nation.

Authoritarian regimes dominated South Korea from 1953 until 1988, although there were periods of rupture. Syngman Rhee's increasingly authoritarian rule from 1953 to 1960 ended with popular uprisings, with students playing a central role. But a new government elected in 1960 was soon overthrown by a military coup led by Park Chung-hee. The South Korean army was large and powerful, with over 600,000 troops, in part supported by the United States, which continued to see South Korea as a bulwark against communism. However, the easy anticommunist rhetoric that Rhee employed so effectively was, as time passed, less able to paper over the fundamental lack of democratic governance. The Park regime actively promoted national economic growth, but the heavy reliance on economic expansion came undone as a slowdown in export-led growth in the early 1970s led to social unrest that in turn was met with restrictions on dissent. The Yusin Constitution of 1972 enshrined a more dictatorial system. Meanwhile economic growth continued at a hectic pace, embodied in about fifty major conglomerates (*chaebol*) whose financial strength was in part a function of their political connections.

Fiscal growth without political participation created disaffected groups. Students were again at the forefront of resistance, and a growing union movement added industrial muscle. Rising antigovernment feeling was tragically enacted in 1979 with Park's assassination. Another military junta soon seized power in bloody confrontations, remaining in power from 1981 to 1987. But the regime could no longer use economic growth to trump the lack of political development. Against a background of fading legitimacy, it announced in 1987 that the president would be chosen through direct elections. The two opposition leaders split the vote, giving the presidency to a military man, Roh Tae-woo. Under the next president, Kim Young-sam, Roh and other members of the military junta were convicted of treason and corruption. The days of the military junta were at an end. South Korean politics, while divisive and lively, with many accusations of corruption and a tendency to authoritarianism, now have a profile more similar to the mature democracies around the world.

While political development has been relatively slow—with South Korea emerging as something resembling a democratic system only in 1987—economic development has been more spectacular. From 1960 to 1990, it was the second fastest-growing economy in the world, driven by export-led growth. South Korea is a manufacturing powerhouse, producing goods sold around the world—from cars and appliances to electronics and computers. From a weak and impoverished agricultural society at the beginning of the twentieth century, the country has developed into a thriving urban industrial nation with an expanding and prosperous middle class. Its decisive insertion into the global economy does, however, make it very susceptible to global cycles of growth and decline. Economic crises in 1997–98 and 2008 exposed the country's reliance on widespread borrowing and its heavy dependence on overseas export markets.

As a modern country, South Korea is extensively and intensively mapped. Building on a long cartographic tradition, there are varieties of maps made at different scales. Figure 7.3 is map of the country produced by the National Geographic Institute in 1998 for overseas distribution. Korean cartography is now an integral part of a worldwide discourse of shared standards, similar conventions, and agreed methods, and mapping practices in South Korea are now similar to those of devel-

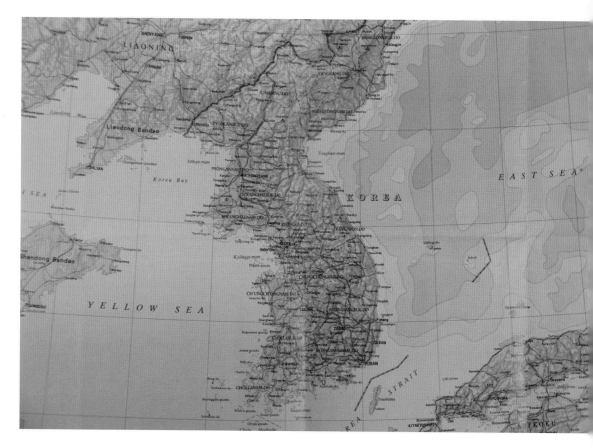

7.3 KOREA, 1998. (American Geographical Society Library, University of Wisconsin–Milwaukee Libraries: 469 D–1998.)

oped countries around the globe. Modernity is reflected and embodied in a standard cartographic language shared with most other countries in the world. The globalization of space involves the adoption of such shared cartographic practices.

In the public sector the South Korean government, specifically the Ministry of Government and Administration and Home Affairs, developed an open source mapping system. Street addresses and building numbers are now combined in a national Geographic Information System that accords with international standards. The Korean Land Corporation is another government-sponsored organization that has a landownership data system, known as the Spatial Information Knowledge System, used in land use planning. A sophisticated and interactive national spatial data infrastructure is used by central and local government agencies as well as by consumers. There is also a vibrant private sector mapmaking and map publishing industry in South Korea. Paju Book City, a specially designed new publishing center

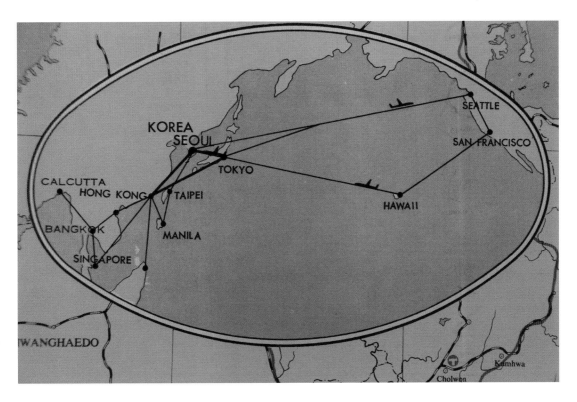

7.4 TOURIST MAP, 1964, South Korean Ministry of Transport. (American Geographical Society Library, University of Wisconsin–Milwaukee Libraries: 469 D-1964.)

northwest of Seoul, includes several map publishing houses. SJ Map, for example, makes a range of cartographic products including school geography texts, globes, atlases, road maps and atlases, and travel maps, as well wall maps of Seoul, Korea, and the world. The use of national and international atlases is an important part of education in South Korea, where cartographic literacy is an integral part of the school curriculum.

How are the many and diverse changes noted above reflected in contemporary maps? I will give two examples. The first way is the growing global reach of the nation's connections. Figure 7.4 is part of a map made for the Ministry of Transportation in 1964, in English and designed for foreign tourists. Superimposed on the territory of South Korea is a diagram showing airline connections to other parts of the world. Compare the limited range of the connections with the global range depicted in figure 7.5, a map made in 2000 by the National Tourist Organization. Air routes now link Seoul to the major cities of the world. The map suggests other ties,

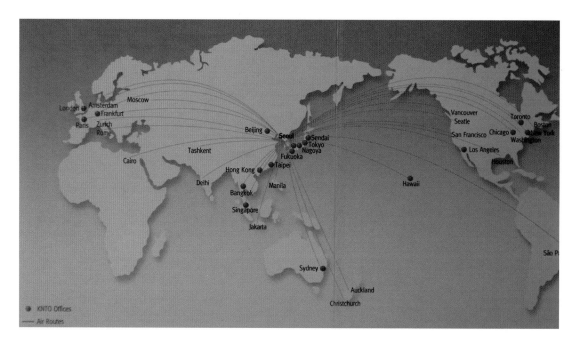

7.5 TOURIST MAP, 2000, South Korean National Tourist Organization. (American Geographical Society Library, University of Wisconsin–Milwaukee Libraries: 469 D-2000.)

including investment and trade in commodities. South Korea is firmly linked into global flows of capital and goods.

Despite the economic globalization, cultural globalization is less well developed. The country retains its ethnic homogeneity and discourages foreign immigration. However, the hosting of the 1988 Olympics Games in Seoul served as an important platform to integrate the country more fully into a global awareness and to foster a more cosmopolitan character. The second example I use, then, is the rise of tourist maps for an international audience. Many of the maps of Korea produced in Korea now have text in both Korean and English. International tourism is encouraged and promoted. Figure 7.6, for example, is a 2002 map of Seoul produced in the popular series of Beetle tourist maps, made in Korea by GeoMarketing Ltd. Produced for both Korean and English readers, the maps, while thoroughly modern in design, also recall the older tradition of Korean cartographic depictions (see fig. 2.9).

The contemporary cartographic representation of Korea, both North and South, highlights two broader themes. The first is that the globalization of space involves incorporating nations and states into a global mapping project. Nations are

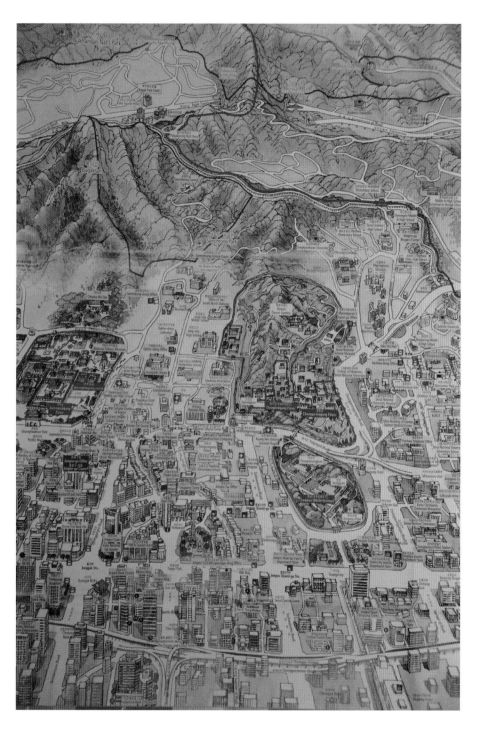

7.6 SEOUL, 2002. (American Geographical Society Library, University of Wisconsin–Milwaukee Libraries: 469. S 46 A-2002.)

mapped by others especially when they are of strategic geopolitical importance, as North and South Korea were during the cold war and remain today. The second is that the mapping of the countries draws on a more universal cartographic language. The distinctiveness of Joseon cartography is replaced by a cartography that, while echoing older national traditions, is in many ways indistinguishable from forms of mapping found in many other countries. The globalization of space involves the growing uniformity of mapping practices across the globe, so that the world is made intelligible to others by a more universal cartographic language. The language of maps is now more widely comprehensible and globally integrated. In the production of global space, cartography is an important medium of a global modernity.

8 CARTROVERSIES

The complex recent history of Korea is reflected in contemporary cartographic depictions. And these depictions, though they adopt a more explicitly scientific and purportedly objective stance, are marked by debate and controversy. A self-consciously scientific mapping cannot solve enduring political conflicts; it can in fact reinforce them by articulating them in maps.

We can identify three of what I term "cartroversies"—debates and tensions in cartographic representation. The first refers to a cartographic silence and confusion regarding Korea's division into North and South. Many contemporary Korean maps do not show the border. Look again at figure 7.3. Korea is presented as one country with no international border and only one capital, Seoul. It is as if the division into North and South had never occurred or the reunification of the peninsula were a contemporary fact. Although there is a "demarcation line" symbol, it does not have the prominence of the symbol for an international political boundary. This is a common feature of contemporary Korean maps—they represent a more unified country than currently exists.

In both North and South Korea there is a cartographic representation of a single country. Especially in school atlases and maps for the general public, the peninsula is still represented as one entity. Korea lives on in the realm of maps, where the division between North and South is regularly ignored and both countries claim sole legitimacy in the peninsula.

The cartographic representation of a country is never innocent of political calculations and aspirations. In the case of Korea it extends to the very definition of country. To take just some contemporary examples: when the *Korea Times*, Seoul's principal English-language daily newspaper, shows a national weather map, it includes the entire peninsula. Even when mapping just North Korea, maps have to situate it in the peninsular context. South Korean maps of North Korea, for example, include an inset showing all of Korea. The division between North and South does not look like an international boundary compared, say, with the boundary with China, but suggests a temporary divide. Prewar provincial boundaries are barely disturbed by the DMZ, which most Korean maps, North and South, narrow to a thin, sometimes invisible line.

Cartographic flexibility in national definition is also revealed in the very latest *National Atlas of Korea,* published in English in 2009 by the National Geographic Information Institute and the Ministry of Land, Transport, and Maritime Affairs of South Korea (Hong 2009). Embossed on the front cover is an outline of the entire peninsula. Inside, the "national" map shows Seoul as the main capital. The atlas has general maps of all the provinces in Korea as well as depicting landforms and geology. But the data on soils, climate, population, and many other socioeconomic and political themes are displayed only for South Korea. The atlas thus represents and lays claim to the entire peninsula as well as depicting the continuing divide. While the Institute and Ministry could assemble a lot of data for South Korea, doing so for North Korea was more difficult. The result is an unusual blend of information for the entire peninsula as well as just for South Korea. Even the maps that show information only for South Korea contain the singular description of "Korea." Maps are never just innocent depictions of territory; they are claims, hopes, and aspirations—in this case the hope for a unified Korea against the stubborn reality of the continuing divide. Each political regime lays claim to territorial legitimacy by representing a peninsular Korea as a single entity with its own capital as the only one.

The lack of a formal peace treaty between the North and South (they did not even both sign the 1953 Armistice Agreement; only North Korea signed along with China and the United States) has led to a complicated and fraught relationship that has shifted over the years. In the wake of the Korean War, the division between the two countries also marked the geopolitical divide in a tense cold war. Tensions eased in the 1980s as the cold war began to thaw. In 1990 the first high-level talks between the two governments took place. Throughout the 1990s there were occasional moves toward reconciliation, but the first sustained effort came with the more conciliatory Sunshine Policy of Kim Dae-jung, who was the South Korean president from 1998 to 2003. The first meeting between leaders of the two countries took place in 2000 between Kim Dae-jung and Kim Jong-il. A second summit occurred in October 2007, between the North Korean leader and a new president, Roh Moo-hyun. But recent missile testing, ship sinkings, and bombardments by North Korea have strained relations, and unification seems a distant prospect. The maps of a single Korean nation-state speak to a hope rather than a reality; they suggest a unified future rather than representing the divided present.

EAST SEA/SEA OF JAPAN

Figure 7.3 also hints at the second cartroversy, the designation of the body of water to the east of Korea. On this map the sea is termed the East Sea. Many non-Korean maps, until comparatively recently, have referred to it as the Sea of Japan.

There is a wider context to this cartroversy. Naming things is how we humanize the world. Naming gives shape and meaning—it turns space into place. Naming is never innocent of politics. The naming of the earth's surface is shaped by the indigenous, the colonial, and the postcolonial populations. The names of places were first given by indigenous peoples, but around the world colonialism meant a rewriting of place. We live in a postcolonial period in which we are more aware of the indigenous legacy and more sensitive to the colonial rewritings.

Naming seas is often more complex than naming land features. Different national territories often surround large bodies of water. The east coast of China faces the west coast of the United States; England's south coast is opposite France's north coast. Seas are shared spaces; there is no simple hegemony over naming as for land surfaces. The larger the sea or ocean, the more nation-states consider themselves to have naming rights, making this a very contentious issue. For very large bodies of water with numerous landmasses and hence nation-states involved, the indigenous names can be so numerous and varied that the colonial names become the standard. Take the large body of water we call the Pacific. It probably had a rich variety of names around its shores as myriad indigenous people named it in their own languages. After the sixteenth century, it was opened up to European trade and mercantile interests. In English it was named with reference to Europe and was originally called the South Sea. When Magellan crossed the ocean in 1520–21, he encountered no storms and named it Mar Pacifico. Since Spain was the dominant global power, the Spanish name displaced all the indigenous names, and in translation it persists to this day, both because of the continuing Spanish legacy in the region and because it was an easy solution to the complexity. The larger the number of divergent indigenous names, the greater the force of a single colonial name.

Territories especially try to influence the naming of seas close to their own coastlines, which are part of their national identity. Conflict is greater where relatively few countries have given a sea indigenous names. Again we can compare the Pacific Ocean, a vast body of water bordered by many original communities and nation-states, with a small sea surrounded by only a few. In the case of the Sea of Japan/East Sea, only Korea, Japan, and Russia, for whom it was at the edge of its empire, surround a relatively small body of water. For most purposes the proximate interests are thus only Japan and Korea. That the two countries have a colonial relationship makes the naming controversy all the more contentious.

Returning to the specific case: the historical background is relatively simple. Until the sixteenth century the Japanese had few maps giving names to the seas far from their immediate shoreline. Japanese cartography was almost entirely concerned with national representation. The Joseon kingdom had a large number of maps employing the term Dong Hae (East Sea), which predates even the Joseon—the name East Sea was inscribed on a monument to King Gwanggaeto erected in 414. The term, however, may be a simple adoption of the Chinese term East Sea, then used to designate all the sea off China's east coast. The early Joseon maps seem to depict the outlying islands of Ulleungdo and Dokdo. The territorial expansion of the Joseon into these islands prompted the wider naming of the sea beyond the narrow coastal fringe. Figure 8.1, for example, is a small map of the country published in a 1530 edition of *Tonguk yoji sungnam* (Augmented Survey of the Geography of Korea), first compiled in 1481, printed in 1487, and reprinted in 1499 and 1530. Titled *Paldo chongdo* (Map of Eight Provinces), the map was deliberately crude in case it fell into foreign hands and because its main purpose was to highlight rivers and mountains used in rituals. The map seems to note the two islands of Ulleungdo and Dokdo, although they appear to be reversed. The designation East Sea is marked directly west of these two islands. The map is both diagrammatic and pictorial: notice the intricate depiction of waves. This map shows the very early use of the term East Sea. Because the Joseon had de jure control over the islands, they had a name for the sea surrounding them: the East Sea. The formal control of the Japanese, in contrast, did not extend very far from their shores, so the more distant sea was seen as more of a blank space. While they sometimes called the sea close to their coast a variant of the Sea of Japan, many of their maps until the late nineteenth century used Sea of Joseon to refer to the sea between Japan and Korea.

Western explorers, merchants, and missionaries used a variety of names. They often called it Oriental Sea, an echo of the term Ptolemy used more than fifteen hundred years earlier. Other names included Sea of Korea (including disparate spellings such as Corea and Corée), Eastern Sea, Sea of China, and Sea of Japan. The names were applied liberally, often marked in different places on different maps, with little cartographic standardization. Guillaume Delisle's 1705 map of Asia has the dual names Mer Orientale and Mer de Corée. From the sixteenth

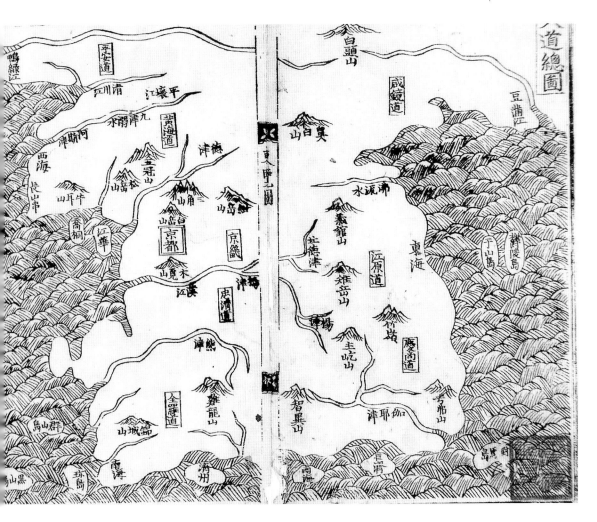

8.1 KOREA, 1530. (Kyujanggak Archives, Seoul National University.)

to the eighteenth century, Sea of Korea was a dominant form. Examples include Robert Dudley's 1646–47 map of Asia, which has Mare di Corai, and Emmanuel Bowen's 1754 map of Asia, which simply has Sea of Korea. A map of Asia published by John Senex in London in 1720 clearly uses the term Eastern Sea or Corea Sea. This map is acutely aware of Japanese territory, since the Japanese island of Oki (shown on the map as Oqui) is prominently displayed, yet the sea is still termed Eastern Sea or Corea Sea. There were variations: the Dutch, having greater contact with the Japanese, tended to use Sea of Japan more often than other European powers, especially the English and then the British. A chart of the northwest

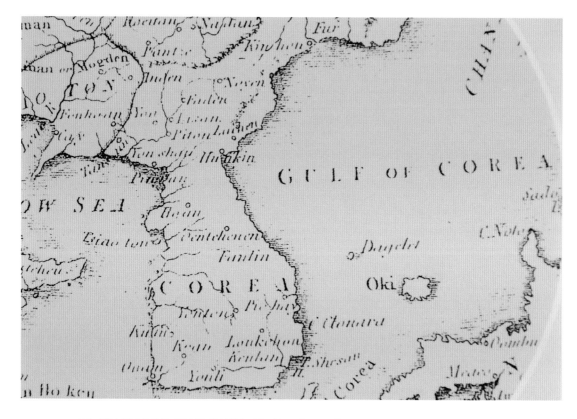

8.2 CHART OF THE COAST OF ASIA, 1794. (Henry Roberts, *Chart of the N.W. Coast of America and the N.E. Coast of Asia* [London: William Faden, 1794]. Courtesy of the Library of Congress, Washington, DC.)

coast of America and northeast coast of Asia made by Captain Cook in 1778–79 and published in London in 1794 by William Faden uses the term Gulf of Corea. This published map also highlights the Korean island of Dagelet (Ulleungdo) and the Japanese island of Oki (see fig. 8.2). A British map made by W. H. Allen in London in 1844 used the descriptor Tong-Hae or Eastern Sea. The range of usages can be found in the Sea of Korea Map Collection in the Library of the University of Southern California. The 172 maps are available online at http://digitallibrary.usc.edu/search/controller/collection/seakorea-m1.html.

Japanese maps include few names for the sea until the end of the eighteenth century. Japan's cartographic parochialism meant there were few depictions of areas outside the national territory. In the nineteenth century many Japanese maps used variants of Sea of Joseon, such as the 1809 map of Japan and the 1810 *Shintei bankoku zenzu* by Takahashi Kageyasu, the 1870 *Bankoku chikyu bunzu* by Hashimoto Gyokyuransai, and the 1885 globe by Horiuchi Naotada. While some maps used just Sea of Japan, such as the 1858 *Yochi kokai zu* by Takeda Kango, many Japanese maps used both Sea of Japan and Sea of Joseon, such as the 1838 map *Yochi zenzu* by Kurihara Shincho and the 1870 *Dai nihon shishin zenzu* by Hashimoto Gyokuransai. A common feature of many Japanese maps until the end of the nineteenth century was to name the larger sea area between the two countries Sea of Joseon while using Sea of Japan for the area to the east of the Japanese coastline. The Japanese mapmaker Kurihara Nobuaki produced a double hemisphere map of the world in 1848 that designated the sea between the two countries as Sea of Korea. Sea of Great Japan was used to refer to the Pacific Ocean. However, by the end of the nineteenth century, when Japanese imperialism and militarism were in full stride, all Japanese maps and globes used Sea of Japan for the sea between Japan and Korea. The last Japanese map to employ the term East Sea was published in 1894. After Japan's victory in the Sino-Japanese War of 1894, Japanese maps initially displaced East Sea to the East China Sea, eventually dropping the term altogether and using only Sea of Japan for the sea between Japan and Korea. The Korean name was erased and replaced by Sea of Japan. Japan's annexation of Korea in 1910 solidified the new naming.

By the end of the nineteenth century, Sea of Japan was also used in European and American maps and atlases, displacing Sea of Korea. Japan was entering the international scene as the dominant regional geopolitical power, and that power included international recognition and legitimacy for its names for the surrounding seas. Colton's 1885 map of Asia, for example, uses Japan Sea, as does an 1895 Rand McNally map of Asia.

With the Japanese takeover of Korea, the sea was singularly named by and after the colonial power. Korea was under Japanese domination when the 1929 Monaco Conference of the International Hydrographic Organization resulted in the publication of the first edition of the authoritative *Limits of Oceans and Seas*, which used Sea of Japan. The Sea of Japan designation persisted after the end of Japanese colonial rule in 1945 because of the standardizing legacy from the 1929 conference. But there was another reason too. In the immediate aftermath of World War II, the American mapmakers drew heavily from Japanese maps of Korea, including place-names. Figure 8.3 shows the source credits for a 1946 United States map of part of Korea. The Japanese influence is obvious. In the small sliver of the map shown, Japanese names are clearly visible beside every Korean name. Figure 8.4 is a 1951 American grid map of Korea at the scale of 1:250,000. The map extends into the sea east of Korea, where the direct adoption of the Japanese names is apparent. The United States produced both military and commercial maps of Korea and the wider region, in a context saturated by Japanese maps and Japanese names. While

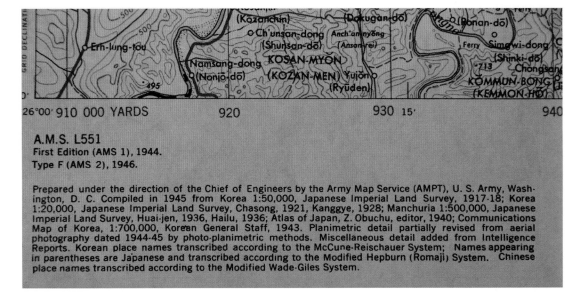

A.M.S. L551
First Edition (AMS 1), 1944.
Type F (AMS 2), 1946.

Prepared under the direction of the Chief of Engineers by the Army Map Service (AMPT), U. S. Army, Washington, D. C. Compiled in 1945 from Korea 1:50,000, Japanese Imperial Land Survey, 1917-18; Korea 1:20,000, Japanese Imperial Land Survey, Chasong, 1921, Kanggye, 1928; Manchuria 1:500,000, Japanese Imperial Land Survey, Huai-jen, 1936, Hailu, 1936; Atlas of Japan, Z. Obuchu, editor, 1940; Communications Map of Korea, 1:700,000, Korean General Staff, 1943. Planimetric detail partially revised from aerial photography dated 1944-45 by photo-planimetric methods. Miscellaneous detail added from Intelligence Reports. Korean place names transcribed according to the McCune-Reischauer System; Names appearing in parentheses are Japanese and transcribed according to the Modified Hepburn (Romaji) System. Chinese place names transcribed according to the Modified Wade-Giles System.

8.3   MAP CREDITS, 1946. (American Geographical Society Library, University of Wisconsin–Milwaukee Libraries: 469 D-1946.)

the place-names within Korea quickly reverted to their Korean originals, the name Sea of Japan persisted, strengthened by Japanese dominance in the late nineteenth century, embodied in the 1929 international agreements, and reinforced through the Japanese cartographic legacy in American postwar maps of Korea.

In summary, while for centuries the Koreans have called the sea the East Sea, it was only in the late nineteenth century that the term Sea of Japan was used. It was not indigenous to the area but was a European name later adopted by the Japanese, often used instead of and sometimes alongside Sea of Korea. In an act of colonial possession, Japan was able to rename it so, and this name was standardized by international agreements drawn up while Japan held power over Korea and reinforced by the American adoption of Japanese place-names.

In 1992, at the sixth Conference on the Standardization of Geographical Names, held in the United Nations headquarters in New York, Korean government officials raised the issue of the East Sea. South Korea became a member of the United Nations only in 1991. In 2007 the two Korean states both made presentations to the ninth United Nations Conference on the Standardization of Geographical Names. North Korea requested that the name East Sea of Korea be used. North Korean maps use East Sea of Korea as well as the designation West Sea of Korea for the

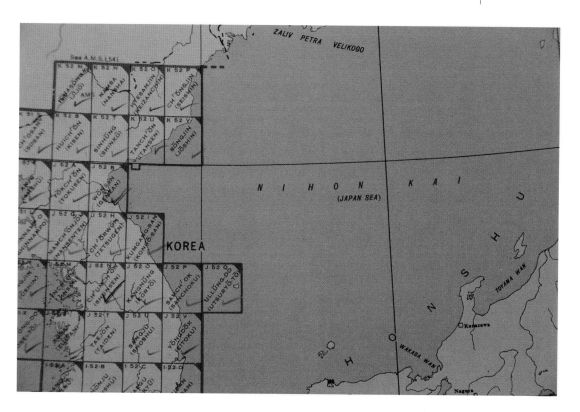

8.4 INDEX MAP, 1951. (American Geographical Society Library, University of Wisconsin–Milwaukee Libraries: 469 D-1951.)

area generally known as the Yellow Sea. The South Korean request, and the official government position, is for a dual system of East Sea / Sea of Japan.

The dual naming is a postcolonial solution compared with the hegemonic and colonial Sea of Japan. A postmodern flexibility of dual naming is now more common around the world; witness the growing use on maps of English Channel / La Manche to refer to the strip of water between southern England and northern France. A joint naming of East Sea / Sea of Japan seems an appropriate, fair, and historically accurate solution and also echoes precolonial namings. In his 1750 map *L'empire du Japon*, Robert de Vaugondy, mapmaker to the king of France, names the sea off the east coast of Korea Mer de Corée and the sea adjacent to Japan Mer du Japon (see fig. 8.4).

The Korean campaign to have dual naming adopted as an international standard takes many forms, from direct government pronouncements and lobbying to

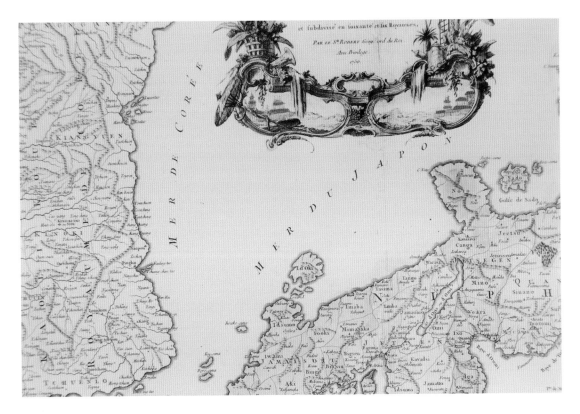

8.5   JAPAN AND KOREA, 1750. (Robert de Vaugondy, *L'empire du Japon* [Paris: Gilles, 1750]. Courtesy of the Library of Congress, Washington, DC.)

various social movements, which often demand a singular naming of East Sea. One group, www.forthenextgeneration.com, regularly takes out advertisements accompanied by maps to correct what they see as the error of designating the sea as the Sea of Japan. The organization published a full-page ad in the *New York Times* on May 11, 2009, and in the *Washington Post* on August 12, 2009, complete with maps, to dispute the use of Sea of Japan and claim legitimacy for East Sea. The campaigns have been successful in shifting global public opinion. *National Geographic*, for example, now uses the dual designation Sea of Japan / East Sea on all its maps. Increasingly, dual naming is used on many non-Japanese maps of the region.

We live in a postcolonial world, and we need a postcolonial sensitivity. Around the world there is recognition of the evils of the past. The Australian government has formally apologized to the indigenous peoples; the United States House of Representatives as well as numerous state legislatures around the country have passed resolutions apologizing for the inhumanity of slavery. Worldwide, communities and nations are responding to colonial mentalities rather than denying or perpetuating them. Global citizenship now implies—indeed demands—an honest historical reckoning of a nation's colonial past. And to move into the future as a proactive force in the global community, a nation, and especially its leaders, needs to see the colonial legacy that continues to guide its policies. By recognizing the dual naming of the East Sea/Sea of Japan, Japan can invoke a postcolonial sensitivity and embrace a more effective global citizenship. Japan needs to accept its colonial legacy in order to transcend it and become a more effective and morally powerful force in the world.

DOKDO

The third cartroversy is the representation of Dokdo on maps. If we look again at figure 7.3, we can see a small red line in the middle of what is designated as the East Sea, encompassing Dokdo within the space of Korean national sovereignty. The area is administered by South Korea but claimed by Japan. Figure 8.6 is a closeup of the islands from a map in the 2007 *Ocean Atlas of Korea*.

Dokdo, which the Japanese refer to as Takeshima, consists of two small rocky islands surrounded by approximately thirty-three smaller rocks. In total it amounts to just under two square kilometers. For such a small place, it has generated intense political heat. Part of the problem lies in geography. Dokdo is far from both Korea and Japan. For centuries both countries discouraged their citizens from sailing too far from the coast, and it was always on the edge of the effective range of control. The old Korean maps often show two islands off the east coast of the country. Look carefully again at figure 5.14, and it is clear that two islands are shown there. One interpretation is that they are the two island groups of Ulleungdo and Dokdo. If this is true, then Dokdo and Ulleungdo were incorporated into the Silla kingdom in 512 as one unit, Usanguk. They were annexed by the Goreyo regime in 930 and became part of the Joseon kingdom. They are clearly shown in the early Joseon map of 1530 in figure 8.1. Although they were part of Joseon territory, they were always on the farthest reaches of control, subject to disputes and to pirate attacks. From 1416 to 1881 the Joseon adopted a vacant island policy and did not promote settlement in the region. However, a dispute between Korean and Japanese fishermen in the late seventeenth century led to official Japanese interest. Officials of the Tokugawa shogunate decided in 1696 and again in 1699 that the islands were part of Korea, essentially ceding them to the Koreans.

Another interpretation, strong among those who back Japan's claims, is that the two islands shown off the east coast on old Korean maps are in fact Ulleungdo

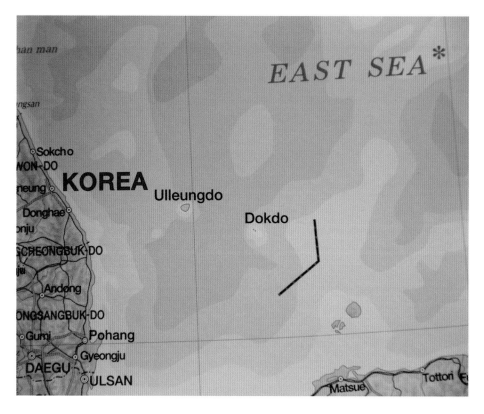

8.6 EAST SEA. (*Ocean Atlas of Korea* [Republic of Korea: National Oceanographic Research Institute, 2007].)

and its smaller eastern neighbor the island of Judo, not Dokdo. All the discussion above then refers just to Ulleungdo, not to Ulleungdo and Dokdo. Some argue that the two islands are so far apart that their closeness on the old maps is a confusion between Dokdo and Jokdo. However, this is not a sound argument. As I have shown throughout this book, Joseon cartographers were sophisticated enough to use relative as well as absolute space in their maps. The islands' closeness on old maps is simply a result of placing them close to the coastline to highlight territorial spatial relations rather than a depiction of Euclidean space.

For the Koreans, their sovereignty over the island of Dokdo was clear and unambiguous. For the imperial Japanese, in contrast, Dokdo was *terra nullius*. The expansionist regime of 1868 in Japan adopted a new, more aggressive territorial attitude. In 1905 a local government in Japan, Shimane Prefecture, unilaterally incorporated Dokdo/Takeshima into its territory. The Japanese asserted that the island was unclaimed and, even if not, had always been part of Japan. There are reports from

Korean sources in the late seventeenth and late nineteenth century of Japanese loggers and fishermen in the islands. Yet Dokdo was under Korean sovereignty until 1905, when it was annexed as part of Japanese imperial expansion. The annexation was always something of a risk. Japanese officials of the time wondered if the act, with minimal gains, would arouse international suspicion about Japan's intentions toward Korea. Other officials argued that in the light of deteriorating international conditions, it was important to establish a base to keep watch on foreign ships. The decision to annex, then, was less about enacting a historical claim than a calculated contemporary geopolitical power play. After Japan's formal takeover of Korea in 1910, the question of de jure ownership became moot, since the island came under Japanese control along with all of Korea.

With Japan's defeat by the Allied forces in 1945, its title to territory through colonial domination was effectively nullified. Allied powers specifically excluded the islands from Japanese control in 1946. It is here that things get a bit hazy. Japan managed to influence the San Francisco Peace Treaty so that the islands' sovereignty was put in doubt. Successive drafts of the treaty had conflicting conclusions. Korea's case was weakened by President Syngman Rhee's failure to make the Korean case for Dokdo. He focused instead on the quixotic case of Korean sovereignty over Tsushima Island. To reinforce their claim, the South Koreans established a lighthouse on Dokdo in 1954 and have effectively controlled the place since then. In 1982 the island was designated a National Cultural Heritage–Natural Monument, and in 2003 the island was assigned a South Korean postal code. Today Dokdo is under effective Korean control.

Even as Japan-Korea relations improve, Dokdo remains unresolved. It is used as a rallying platform for nationalist Japanese politicians, especially those wishing to burnish an image of tough nationalism. When Japan's Shimane Prefecture declared a Takeshima Day in 2005, it provoked mass rallies and demonstrations in Korea. While both Japan and South Korea claim the island, the de facto reality is that Dokdo remains under effective South Korean control, and Japan is unlikely to invade. The issue is not just one of political posturing. The rich fishing stocks and the existence of valuable gas hydrates makes the competing claims economically relevant. Even the smallest islands can become opportunities to extend two-hundred-mile exclusive economic zones.

Dokdo is located in a difficult space on the extreme edge of both states, 135 miles from mainland Korea and 150 miles from Japan. The nearest islands are the Japanese island of Oki, 98 miles away, and the Korean island of Ulleungdo, just 54 miles away. Largely unpopulated throughout most of its history, even today Dokdo has only two permanent residents, an octopus fisherman and his wife. Dodko also occupies a complex cartographic space. The Korean claim is backed up with reference to old maps and documents that seem to show Korea's de jure, if not de facto, possession. However, the confusion on many maps and documents between Dokdo and the island of Ulleungdo and its more immediate neighbors makes the cartographic record less than watertight. In summary, Korean maps show indications of Dokdo as Korean, while the Japanese cartographic record either tends to confirm Korean sovereignty or fails even to show Dokdo.

The island also occupies a complex discursive space between Japan and Korea. Japan's claim to the island is based on its colonial expansionist era, not exactly a strong point in a more postcolonial world. For Japan, then, the ownership of Dokdo and the East Sea/Sea of Japan naming controversy are lingering colonial legacies that have morphed into an issue of national prestige. In both cases the tide is against them; there is growing use of the dual designation East Sea/Sea of Japan, and Dokdo is firmly in Korean possession. For the Koreans, the naming controversy and Dokdo represent remnants of a colonial history they want to transcend. In a reversal of the global awareness of the late nineteenth century, the Koreans are exerting more influence on global opinion about the naming of East Sea/Sea of Japan as well as about sovereignty over Dokdo. This is evident even in an examination of world atlases recently produced by various publishers around the world, including Dorling Kindersley, Gallimard, HarperCollins, National Geographic, Pearson, Philip's, Rand McNally, the Times, Oxford University Press, and Westermann, which have all adopted a dual naming of East Sea/Sea of Japan.

What is interesting in the current cartroversies is the continual heavy use of old maps as historical documents. Both the Korean and the Japanese claims are filled with reference to old maps and to mapmakers long dead. Punch in either name into an Internet search engine and you will be amazed at the important role the history of cartography plays in the discussion. Old atlases and ancient charts are rolled out to buttress one claim or another. There is, however, a presupposition that maps are timeless, fixed in meaning and reliable in their claims. But maps are unreliable witnesses. Names on maps are best considered mobile, contested, hybrid things that constantly change meaning as well as location over the years. They are not immutable descriptors that are permanently anchored in singular meanings or even fixed in location. They are too fluid to make strong cases. Let me end by repeating what I wrote at the beginning of the book.

Maps are not simply technical constructs; they are documents that tell us much about society. Maps have multiple meanings: they serve decorative purposes as well as practical applications, and they have symbolic importance and ideological underpinnings. They are pictures of the world that embody technical progress, social development, and political conflict. Maps are complex things—material objects and social documents, technical texts and practical devices. More than just depictions of territory, they are political statements, social arguments, and discourses both subtle and simple. Maps are complicated texts used for a variety of purposes and read by diverse readers.

In complex historical controversies maps are suggestive guides, part of the invention of tradition and the production of the future, aids to flexible and creative thinking. They are not revealed truth.

This list is limited to works in English.

STANDARD WORKS ON JOSEON CARTOGRAPHY

Han, Y.-W., H.-J. Ahn, and W.-S. Bae. 2008. *The Artistry of Early Korean Cartography*. Larkspur, CA: Tamil Vista Publications.

Ledyard, G. 1994. "Cartography in Korea." In *The History of Cartography,* vol. 2, bk. 2, edited by J. B. Harley and D. Woodward, 235–345. Chicago: University of Chicago Press.

OTHER STUDIES OF KOREAN CARTOGRAPHY

Bae, W.-S. 2008. "Joseon Maps and East Asia." *Korea Journal* 48:46–78.

Choe, S. 1998. "Tokdo in Old Maps." *Korea Observer* 29:187–203.

Chong, H. 1973. "Kim Chong-ho's Map of Korea." *Korea Journal* 13:37–42.

Daniels, M. 1977. "Maps and Cartography in Ancient Korea." *Korea Journal* 17:59–63.

Han, Y. 1995. "The Historical Background of the Production of Old Maps." *Seoul Journal of Korean Studies* 8:69–84.

Hesselink, N. 2004. "Landscape and Soundscape: Geomantic Spatial Mapping in Korean Traditional Music." *Journal of Musicological Research* 23:265–88.

Hur, Y. 1990. "Choson Dynasty Maps of Seoul." *Korea Journal* 30:21–35.

Kane, D. 2003. "The Inscrutable Father of Korean Cartography." *Exploring Mercator's World* 8:30–37.

Jeon, S. 1974. *Science and Technology in Korea: Traditional Instruments and Technologies*. Cambridge, MA: MIT Press.

Kim, S. 2000. "A Study on the Historical Geography of East Sea." *Journal of Business History* 1:43–56.

Ledyard, G. 1991. "The Kangnido: A Korean World Map, 1402." In *Circa 1492: Art in the Age of Exploration*, edited by Jay A. Levenson. New Haven, CT: Yale University Press.

Lee, C. 1972. "Old Maps of Korea: Historical Sketch." *Korea Journal* 12:4–14.

———. (1977) 1991. *Old Maps of Korea*. Seoul: Korean Library Science Research.

Lee, K., S. Kim, and Jung-chul Soh. 2002. *East Sea in World Maps*. Seoul: Society for East Sea.

Mackay, A. L. 1975. "Kim Su-hong and the Korean Cartographic Tradition." *Imago Mundi* 27:27–38.

McCune, S. 1977. "World Maps by Korean Cartographers." *Journal of Social Sciences and Humanities* 45:1–8.

———. 1980. "The Korean Cartographic Tradition, Its Cross-Cultural Relations." In *Papers of the First International Conference on Korean Studies*, 724–70. Seoul: Academy of Korean Studies.

———. 1983. "Korean Maps of the Yi Dynasty." *Korean Culture* 4:21–31.

———. 1990. "The Chonha Do: A Korean World Map." *Journal of Modern Korean Studies* 4:1–8.

Park, C.-S. 2008. "Mapping the World: The Joseon Worldview as Seen through Old Maps. *Korea Journal* 48:5–7.

Robinson, K. R. 2007. "Choson Korea in the Ryukoku Kangnido: Dating the Oldest Extant Korean Map of the World (15th Century)." *Imago Mundi* 59:177–92.

Savenije, H. 2000. "Korea in Western Cartography." *Korean Culture* 21:4–19.

Society for East Sea. 2004. *East Sea in Old Western Maps with Emphasis on the 17–18th Centuries*. Seoul: Society for East Sea and Korean Overseas Information Service.

Stephenson, F. R. 1994. "Chinese and Korean Star Maps and Catalogs." In *The History of Cartography*, vol. 2, bk. 2, *Cartography in the Traditional East and Southeast Asian Societies*, edited by J. B. Harley and D. Woodward. Chicago: University of Chicago Press.

Thrower, N., and Y. Kim. 1967. "Dong-Kook-Yu-Ji-Do: A Recently Discovered Manuscript of a Map of Korea." *Imago Mundi* 21:30–49.

Yi, P. 1960. "The Impact of the Western World on Korea in the 19th Century." *Journal of World History* 5:957–74.

Yoon, H. 1992. "The Traditional Standard Korean Maps and Geomancy." *New Zealand Map Society Journal* 6:3–9.

STUDIES OF ASIAN CARTOGRAPHY AND MAPS THAT REFERENCE KOREA

Harley, J. B. and D. Woodward, eds. 1994. *The History of Cartography*. Vol. 2, bk. 2, *Cartography in the Traditional East and Southeast Asian Societies*. Chicago: University of Chicago Press.

Needham, J., and W. Ling. 1959. *Science and Civilization in China*. Vol. 3, *Mathematics and Sciences of the Heavens and the Earth*. Cambridge: Cambridge University Press.

Smith, R. J. 1996. *Chinese Maps: Images of "All under Heaven."* Oxford: Oxford University Press.

Wallis, H. 1965. "The Influence of Father Ricci on Far Eastern Cartography." *Imago Mundi* 19:38–45.

WORK COVERING CRITICAL GEOGRAPHIES OF KOREA

Tangherlini, T. R., and S. Yea, eds. 2008. *Sitings: Critical Approaches to Korean Geography.* Honolulu: University of Hawaii Press.

WEBSITES CONTAINING IMAGES OF OLD MAPS OF KOREA

Korea as seen through Western cartography: http://www.cartography.henny-savenije .pe.kr/.

Old maps of Asia: http://www.maphistory.info/imageasia.html.

A selection of Korean atlases: http://memory.loc.gov/cgibin/query/r?ammem/gmd:@ FIELD(SUBJ+@band(+Korea++Maps++Early+works+to+1800++)). The University of Southern California Library Collection of maps from 1606 to 1895 that focuses on the East Sea/Sea of Japan issue: http://digitallibrary.usc.edu/search/controller/collection/ seakorea-m1.html. Japanese perspectives: http://dbs.library.tohoku.ac.jp/gaihozu/.

There are a number of websites devoted to the Korean War. The US Army Center of Military History has a good collection of war maps: http://www.history.army.mil/ BOOKS/MAPS.HTM.

For more recent cartographic images of North and South Korea see: http://www.nkecon watch.com/north-korea-uncovered-google-earth/ and also http://atlas.ngii.go.kr/ english/.

Agata, A. N. 2005. "American Missionaries in Korea and U.S.–Japan Relations." *Japanese Journal of American Studies* 16:159–78.

Ahn, H.-J. 2008. "Early Cartography and Painting." In *The Artistry of Early Korean Cartography*, edited by Y.-W. Han, H.-J. Ahn, and W.-S. Bae, 133–55. Larkspur, CA: Tamil Vista Publications.

Bae, W.-S. 2008. "Joseon Maps and East Asia." *Korea Journal* 48:46–78.

Bagrow, L. 1958. *The Atlas of Siberia by Semyon Remezov*. The Hague: Mouton.

Baker, D. 2003. *Living Dangerously in Korea: The Western Experience, 1900–1950*. Norwalk, CT: Eastbridge.

Bassin, M. 1999. *Imperial Visions: Nationalist Imagination and Geographical Expansion in the Russian Far East, 1840–1865*. Cambridge: Cambridge University Press.

Bergreen, L. 2003. *Over the Edge of the World*. New York: Morrow.

Berry, M. E. 2007. *Japan in Print: Information and Nation in the Early Modern Period*. Berkeley: University of California Press.

Blusse, L. 2008. *Visible Cities: Canton, Nagasaki and Batavia and the Coming of the Americans*. Cambridge, MA: Harvard University Press.

Broughton, W. 1804. *A Voyage of Discovery to the North Pacific Ocean*. London: Cadell and Davis.

Chang, M. M. 2003. *China in European Maps*. Hong Kong: Hong Kong University of Science and Technology Library.

Cheong, S., and L. Kihan. 2000. "A Study of 16th-Century Western Books on Korea: The Birth of an Image." *Korea Journal* 40:244–83.

Choe, Y. 2005. *Land and Life: A Historical Geographical Exploration of Korea*. Fremont, CA: Jain.

Chong, H. 1973. Kim Chong-ho's Map of Korea. *Korea Journal* 13:37–42.

Dalrymple, A. 1769. *A Plan for Extending the Commerce of This Kingdom and of the East-India-Company*. London: J. Nourse and T. Payne.

De Medina, J. G. R. 1991. *The Catholic Church in Korea: Its Origins, 1566–1784.* Rome: Istituto Storico.

Dunch, R. 2002. "Beyond Cultural Imperialism: Cultural Theory, Christian Missions and Global Modernity." *History and Theory* 39:301–25.

Dunmore, J., ed. 1994. *The Journal of Jean-François de Galaup, Comte de La Pérouse, 1785–1788.* London: Haklyut Society.

Fry, T. H. 1973. "The Commercial Ambitions behind Captain Cook's Last Voyage." *New Zealand Journal of History* 7:186–91.

Haboush, J. K. 2009. "Creating a Society of Civic Culture: The Early Joseon, 1392–1592." In *Art of the Korean Renaissance, 1400–1600*, edited by S. Lee, 1–17. New Haven, CT: Yale University Press and Metropolitan Museum of Art.

Han, Y.-W. 2008. "A Korean Map in the Possession of the National Library of France." In *The Artistry of Early Korean Cartography*, edited by Y.-W. Han, H.-J. Ahn, and B.-W. Sung, 159–83. Larkspur, CA: Tamil Vista Publications.

Han, Y.-W., H.-J. Ahn, and B.-W. Sung, eds. 2008. *The Artistry of Early Korean Cartography.* Larkspur, CA: Tamil Vista Publications.

Harley, J. B. 2002. *The New Nature of Maps: Essays in the History of Cartography.* Baltimore: Johns Hopkins University Press.

Hong, G.-B. 2009. *The National Atlas of Korea.* Suwon, South Korea: National Geographic Information Institute/Ministry of Land, Transport and Maritime Affairs.

Hostetler, L. 2001. *Qing Colonial Enterprise: Ethnography and Cartography in Early Modern China.* Chicago: University of Chicago Press.

———. 2009. Contending Cartographic Claims? The Qing Empire in Manchu, Chinese and European Maps. In *The Imperial Map: Cartography and the Mastery of Empire*, edited by James R. Ackerman, 93–132. Chicago: University of Chicago Press.

Jeon, J. H. 2008. "Spatial Consciousness Represented in Provincial Maps from the Late Joseon Period. *Korea Journal* 48:106–34.

Kang, S. 2008. Frontier Maps from the Late Joseon Period and the Joseon People's Perceptions of the Northern Territory. *Korea Journal* 48:80–105.

Kivelson, V. 2006. *Cartographies of Tsardom: The Land and Its Meaning in Seventeenth-Century Russia.* Ithaca, NY: Cornell University Press.

Ledyard, G. 1994. "Cartography in Korea." In *The History of Cartography*, vol. 2, bk. 2, *Cartography in the Traditional East and Southeast Asian Societies*, edited by J. B. Harley and D. Woodward, 235–345. Chicago: University of Chicago Press.

Lee, C. (1977) 1991. *Old Maps of Korea.* Seoul: Korean Library Science Research.

Livingstone, D., and C. Withers, eds. 2000. *Geography and Enlightenment.* Chicago: University of Chicago Press.

Mackay, A. L., et al. 1975. "Kim Su-hong and the Korean Cartographic Tradition." *Imago Mundi* 27:27–38.

McCune, S. 1978. "Old Korean Hand Atlases." *Map Collector*, September, 31–36.

———. 1982. *Annotated Catalogue of Korean Atlases and Maps in the Library of Congress.* Washington, DC: Library of Congress.

Milton, G. 1999. *Nathaniel's Nutmeg, or The True and Incredible Adventures of the Spice Trader Who Changed the Course of History*. New York: Farrar, Straus and Giroux.

Nakamura, N. 1947. "Old Chinese Maps Preserved by the Koreans." *Imago Mundi* 4:3–22.

Needham, J. 1959. *Science and Civilization in China*. Vol. 3, *Mathematics and Sciences of the Heavens and the Earth*. Cambridge: Cambridge University Press.

Oh, S. H. 2008. "Circular World Maps of the Joseon Dynasty: Their Characteristics and World View." *Korea Journal* 48:8–45.

Olshin, B. B. 1996. "A Sixteenth Century Portuguese Report concerning an Early Javanese World Map." *Historia, Ciencias, Saude-Manguinbos* 2:97–104.

Parker, C. H. 2010. *Global Interactions in the Early Modern Age (1400–1800)*. Cambridge: Cambridge University Press.

Potter, S. 2001. "The Elusive Concept of 'Map': Semantic Insights into the Cartographic Heritage of Japan." *Geographical Review of Japan* 74:1–14.

Raj, K. 2007. *Relocating Modern Science: Circulation and the Construction of Knowledge in South Asia and Europe, 1650–1900*. New York: Palgrave Macmillan.

Richter, A. 1952. *Selections from the Notebooks of Leonardo da Vinci*. Oxford: Oxford University Press.

Robinson, K. R. 2007. "Choson Korea in the Ryukoku Kangnido: Dating the Oldest Extant Korean Map of the World (15th Century)." *Imago Mundi* 59:177–92.

Said, E. 1978. *Orientalism*. New York: Pantheon.

Sato, M. 1996. "Imagined Peripheries: The World and Its People in Japanese Cartographic Imagination." *Diogenes* 173:119–45.

Savenije, H. 2009. *Korea through Western Cartographic Eyes*. Available online at http://www.cartography.henny-savenije.pe.kr/index.htm (accessed December 20, 2009).

Schama, S. 1987. *The Embarrassment of Riches: An Interpretation of Dutch Culture in the Golden Age*. New York: Knopf.

Schmid, A. 2000. "Looking North toward Manchuria." *South Atlantic Quarterly* 99:219–40.

———. 2002. *Korea between Empires, 1895–1919*. New York: Columbia University Press.

Scott, J. C. 1999. *Seeing Like a State: How Certain Schemes to Improve the Human Condition Have Failed*. New Haven, CT: Yale University Press.

Short, J. R. 2000. *Alternative Geographies*. Harlow, UK: Pearson.

———. 2004. *Making Space*. Syracuse, NY: Syracuse University Press.

Short, J. R., and K.-S. Lee. 2010. "Postcolonial Namings in the Immediate Aftermath of the Second World War." Paper presented to Annual Conference of Association of American Geographers. Washington, DC, April 15.

Smith, R. J. 1996. *Chinese Maps: Images of "All under Heaven."* Oxford: Oxford University Press.

Thrower, N., and Y. Kim. 1967. "Dong-Kook-Yu-Ji-Do: A Recently Discovered Manuscript of a Map of Korea." *Imago Mundi* 21:30–49.

Tolmacheva, M. 2000. "The Early Russian Exploration and Mapping of the Chinese Frontier." *Cahiers du Monde Russe* 41:41–56.

Tyacke, S. 2008. "Gabriel Tatton's Maritime Atlas of the East Indies, 1620–1621: Portsmouth Royal Naval Museum Admiralty Library Manuscript, MSS 352." *Imago Mundi* 60:39–62.

Walker, B. L. 2007. "Mamiya Rinzo and the Japanese Exploration of Sakhalin Island: Cartography and Empire." *Journal of Historical Geography* 33:283–13.

Walravens, H. 1991. "Father Verbiest's Chinese World Map, 1674." *Imago Mundi* 43:31–47.

Walter, L. ed. 1994. *Japan: A Cartographic Vision*. Munich: Prestel.

Yonemoto, M. 2003. *Mapping Early Modern Japan: Space, Place and Culture in the Tokugawa Period (1603–1868)*. Berkeley: University of California Press.

Young, R. D. 2003. "Treaties, Extraterritorial Rights and American Protestant Missionaries in Late Joseon Period." *Korea Journal* 43:174–203.